Student Workbook

Basic Mathematics Skills

August V. Treff
Donald H. Jacobs

AGS®
American Guidance Service
Circle Pines, Minnesota 55014-1796

Circle Pines, Minnesota 55014

Printed in the United States of America.
ISBN: 0-7916-0061-0
Order Number: 80032

Contents

Name ____________________

Date ______________

PLACE VALUE

► Write the name of the place for each underlined digit.

1) 2,406 tens	2) 128 ______	3) 70,835 ______
4) 17,501 ______	5) 301,339 ______	6) 10,002 ______
7) 491,918 ______	8) 46,023 ______	9) 6,005 ______
10) 59,700 ______	11) 34,000 ______	12) 500,069 ______
13) 341 ______	14) 10,000 ______	15) 1,000,000 ______
16) 3,902,885 ______	17) 503 ______	18) 16,030 ______
19) 2,000,003 ______	20) 73,999 ______	21) 294 ______
22) 5,020,007 ______	23) 919,078 ______	24) 4,009 ______
25) 5,683 ______	26) 687,633 ______	27) 48,040 ______
28) 384,995 ______	29) 8,837 ______	30) 23,000,821 ______
31) 1,010,001 ______	32) 53 ______	33) 5,078 ______
34) 708,583 ______	35) 61,222 ______	36) 701,865 ______
37) 70,738 ______	38) 501,775 ______	39) 102,895 ______
40) 71,990 ______	41) 88,210 ______	42) 90,909 ______
43) 203,872,221 ______	44) 910,573 ______	45) 10,710 ______
46) 737,098 ______	47) 40,910 ______	48) 10,002 ______
49) 9,033,921 ______	50) 810,022,033 ______	51) 300,941 ______
52) 2,671 ______	53) 94,724 ______	54) 803,921 ______
55) 506 ______	56) 1,034 ______	57) 920 ______
58) 1,023 ______	59) 462,987 ______	60) 10,935 ______

Name ______________________

Date ______________

WRITING NUMBERS

► Write the following numerals in words.

1) 1,208 one thousand two hundred eight

2) 204

3) 4,801

4) 80,026

5) 92,224

6) 44,659

7) 602,875

8) 6,096,089

9) 673,218,003

10) 830,002

11) 755,061

12) 8,038,993,000

Name ____________________

Date ____________

NUMBER TRANSLATIONS

► Write the following amounts in numerals.

1) Three thousand, five hundred thirty-six — 3,536
2) Five hundred six ______
3) Seven hundred forty-nine ______
4) Five thousand, nine ______
5) Seven thousand, three hundred twenty-one ______
6) Nine thousand, two ______
7) Nine thousand, five hundred ______
8) Thirty-one thousand, four ______
9) Fifty-seven thousand, nine hundred ______
10) Eighty thousand, six hundred thirty-two ______
11) Forty-two thousand, three ______
12) Ninety-one thousand, four hundred eleven ______
13) Seven hundred thousand ______
14) Nine hundred thousand, sixty-four ______
15) Seven hundred seventy-one thousand, five hundred forty-nine ______
16) Four hundred fifty-five million ______
17) Three hundred five million, twenty-eight thousand, two ______
18) Eight thousand, eleven ______
19) Three hundred sixty-three thousand, five hundred four ______
20) Seventy thousand, nine hundred forty-two ______
21) Forty-seven thousand, eighty-one ______
22) Eleven million, twenty-nine thousand, four hundred ______
23) Two hundred million, three hundred four ______
24) Fourteen thousand, ninety-eight ______
25) Fifty-three thousand one ______

Name ____________________

Date ______________

ROUNDING WHOLE NUMBERS

► Round these numbers to the nearest tens place.

1) 48 = 50	2) 305 = ______	3) 4,056 = ______
4) 408 = ______	5) 9,911 = ______	6) 72,099 = ______
7) 5 = ______	8) 803 = ______	9) 617 = ______
10) 61,092 = ______	11) 777 = ______	12) 290 = ______
13) 4 = ______	14) 18 = ______	15) 102,005 = ______
16) 4,506 = ______	17) 9 = ______	18) 61 = ______

► Round these numbers to the nearest hundreds place.

19) 693 = 700	20) 349 = ______	21) 9,012 = ______
22) 7,521 = ______	23) 43,071 = ______	24) 102,009 = ______
25) 29 = ______	26) 3,002,091 = ______	27) 71 = ______
28) 91,029 = ______	29) 6,018 = ______	30) 33,951 = ______
31) 34,988 = ______	32) 89 = ______	33) 129,999 = ______
34) 10,891 = ______	35) 509 = ______	36) 780 = ______

► Round these numbers to the nearest thousands place.

37) 199 = 0	38) 499 = ______	39) 1,500 = ______
40) 25,509 = ______	41) 999 = ______	42) 90,098 = ______
43) 1,058 = ______	44) 501 = ______	45) 99 = ______
46) 301 = ______	47) 298 = ______	48) 78,475 = ______
49) 470,512 = ______	50) 19,000 = ______	51) 199 = ______
52) 72,055 = ______	53) 1,687 = ______	54) 89 = ______

Example:

```
 11
  18
 162
 171
+  8
 359
```

Name ____________________

Date ____________

ADDITION OF WHOLE NUMBERS

▶ Rewrite the following addends in the vertical form and add.

1) 18 + 162 + 171 + 8 = ________
2) 4 + 37 + 812 + 774 + 1 = ________
3) 8 + 35 + 77 + 273 + 65 = ________
4) 54 + 76 + 90 + 725 = ________
5) 701 + 33 + 83 + 61 + 374 = ________
6) 7 + 837 + 504 + 91 + 522 = ________
7) 93 + 705 + 866 + 73 = ________
8) 45 + 38 + 401 + 5000 = ________
9) 86 + 59 + 63 + 27 + 105 = ________
10) 395 + 57 + 82 + 273 + 88 = ________
11) 304 + 771 + 826 + 776 = ________
12) 366 + 8,261 + 8,837 + 912 = ________
13) 6,372 + 75 + 908 + 76 = ________
14) 874 + 7,601 + 406 + 837 = ________
15) 7,091 + 5,308 + 354 + 34 = ________
16) 645 + 823 + 806 + 7,735 = ________
17) 6,657 + 4,321 + 7,341 = ________
18) 905 + 624 + 861 + 968 = ________
19) 6,241 + 8,548 + 9,092 = ________
20) 558 + 523 + 128 + 8,241 = ________
21) 264 + 63 + 7,253 + 2 = ________
22) 73 + 8,263 + 78 + 521 = ________
23) 42 + 3,547 + 8,142 + 467 = ________
24) 8,263 + 990 + 352 + 37 = ________
25) 7,364 + 364 + 902 + 36 = ________
26) 889 + 902 + 836 + 2,431 = ________
27) 390 + 263 + 7,746 + 477 = ________
28) 3,746 + 7,500 + 9,928 + 388 = ________
29) 6,635 + 809 + 300 + 646 = ________
30) 9,745 + 4,869 + 7,089 + 3,745 = ________
31) 36 + 2,006 + 215 + 116 = ________
32) 2,117 + 3,591 + 6,711 + 2,883 = ________
33) 7,001 + 375 + 6 + 39 = ________
34) 5,613 + 753 + 2,613 + 9,137 = ________

▶ Solve the following word problems with addition.

35) Mark collected 178 pounds of scrap iron and 85 pounds of copper. Find the total weight of the metal. ________

36) Floyd purchased 280 square feet of carpet for his living room and 250 square feet for his bedroom. Find the total number of square feet he purchased. ________

Name ____________________

Date ______________

Example:

$$\begin{array}{r} \overset{7\,12}{\cancel{8}\cancel{2}1} \\ -71 \\ \hline 750 \end{array}$$

SUBTRACTION OF WHOLE NUMBERS

▶ Rewrite these subtraction problems in the vertical form. Then subtract.

1) 821 − 71 = ________ 2) From 384 subtract 75. ________

3) 602 − 113 = ________ 4) From 102 subtract 89. ________

5) 856 − 773 = ________ 6) Subtract 871 from 1029. ________

7) 552 − 498 = ________ 8) Subtract 528 from 717. ________

9) 4,852 − 665 = ________ 10) From 3810 subtract 1922. ________

11) 3,952 − 3,877 = ________ 12) Subtract 9,099 from 10,099. ________

13) 12,923 − 8,973 = ________ 14) From 16,242 subtract 10,987. ________

15) 3,049 − 1,906 = ________ 16) Subtract 786 from 25,004. ________

17) 46,974 − 18,860 = ________ 18) From 65,208 subtract 56,987. ________

19) 7,890 − 5,699 = ________ 20) Subtract 61,098 from 87,987. ________

21) 67,951 − 56,508 = ________ 22) From 10,001 subtract 9,802. ________

23) 78,000 − 6,784 = ________ 24) Subtract 675 from 1000. ________

25) 362,900 − 87,098 = ________ 26) Subtract 81,321 from 601,030. ________

27) 70,981 − 69,673 = ________ 28) From 508,821 subtract 91,055. ________

29) 201,568 − 156,079 = ________ 30) Subtract 82,547 from 90,113. ________

31) 302,711 − 298,035 = ________ 32) Subtract 73,107 from 82,303. ________

▶ Solve the following word problems with subtraction.

33) Leonard sold 185 tickets to the school's Faculty Amateur Show. If he was given 350 tickets to sell, how many did he have left to sell? ________

34) Cassie planned a 475-mile trip. She drove 296 miles the first day. How many miles must she drive the second day to complete her trip? ________

Name ______________________

Date ______________

MULTIPLICATION PRACTICE

▶ Fill in the multiplication facts. Multiply each number in the left column by each number on the top row. Write the product of each pair of numbers in the box where the column and the row meet.

1)

X	0	1	2	3	4	5	6	7	8	9	10
0	0	0	0								
1	0	1	2								
2											
3											
4											
5											
6											
7											
8											
9											
10											

2)

X	6	2	3	0	9	10	8	7	5	4	1
2											
9											
5											
4											
8											
0											
7											
6											
3											
1											
10											

Example:

$$\begin{array}{r} 235 \\ \underline{\times\ 10} \\ 2350 \end{array}$$

Name ______________________

Date ______________

MULTIPLICATION BY POWERS OF TEN

▶ Multiply. Write the answers.

1) 235 × 10 = ________
2) 421 × 100 = ________
3) 4,631 × 10 = ________
4) 6,023 × 100 = ________
5) 702 × 100 = ________
6) 3,011 × 1,000 = ________
7) 3,203 × 100 = ________
8) 26,190 × 10 = ________
9) 1,043 × 100 = ________
10) 50,783 × 1,000 = ________
11) 72 × 1,000 = ________
12) 38 × 1,000 = ________
13) 106 × 100 = ________
14) 81 × 100 = ________
15) 4,123 × 10 = ________
16) 3,007 × 1,000 = ________
17) 962 × 1,000 = ________
18) 300 × 10 = ________
19) 4,305 × 10 = ________
20) 4,020 × 1,000 = ________
21) 412 × 1,000 = ________
22) 906 × 1,000 = ________
23) 10,802 × 100 = ________
24) 104 × 100 = ________
25) 56 × 10 = ________
26) 13 × 100 = ________
27) 9 × 1,000 = ________
28) 83 × 1,000 = ________
29) 183 × 1,000 = ________
30) 7 × 1,000 = ________
31) 801 × 100 = ________
32) 334 × 10 = ________
33) 632 × 10 = ________
34) 4,567 × 100 = ________
35) 5 × 100 = ________
36) 20,304 × 100 = ________
37) 100 × 1,000 = ________
38) 20,011 × 100 = ________
39) 4,302 × 100 = ________
40) 10,001 × 100 = ________
41) 5,562 × 10 = ________
42) 5,563 × 10,000 = ________
43) 1,000 × 2,302 = ________
44) 10,000 × 2,044 = ________
45) 100 × 20,043 = ________
46) 10 × 12,030 = ________
47) 3,800 × 100 = ________
48) 100 × 1,000 = ________

Example:

```
  52
x 42
 104
208
2184
```

Name ____________________

Date ______________

MULTIPLICATION OF WHOLE NUMBERS

▶ Rewrite these multiplication problems in the vertical form and multiply.

1) 52 × 42 = ____________ 2) 61 × 18 = ____________

3) 201 × 43 = ____________ 4) 85 × 72 = ____________

5) 712 × 66 = ____________ 6) 819 × 94 = ____________

7) 465 × 20 = ____________ 8) 762 × 300 = ____________

9) 301 × 300 = ____________ 10) 784 × 100 = ____________

11) 629 × 150 = ____________ 12) 607 × 515 = ____________

13) 5,763 × 501 = ____________ 14) 7,114 × 35 = ____________

15) 920 × 724 = ____________ 16) 856 × 326 = ____________

17) 3,021 × 307 = ____________ 18) 638 × 800 = ____________

19) 4,160 × 110 = ____________ 20) 8,522 × 574 = ____________

21) 5,021 × 4,000 = ____________ 22) 7,000 × 387 = ____________

23) 5,448 × 673 = ____________ 24) 7,361 × 6,000 = ____________

25) 4,000 × 3,000 = ____________ 26) 4,000 × 4,000 = ____________

27) 3,500 × 5,100 = ____________ 28) 6,702 × 1,023 = ____________

29) 6,080 × 6,009 = ____________ 30) 4,455 × 2,021 = ____________

▶ Solve the following word problems with multiplication.

31) Edith jogs 4 miles every day before school for physical fitness. If she jogs 179 days, how many miles will she jog? ________

32) Each student in Mr. Brown's class donated 15 sandwiches to the school picnic. If there were 38 students in Mr. Brown's class, how many sandwiches were donated? ________

Example:

$$\begin{array}{r} 28 \\ 6\overline{)168} \\ \underline{12} \\ 48 \\ \underline{48} \\ 0 \end{array}$$

Name ____________________

Date ______________

DIVISION OF WHOLE NUMBERS

▶ Rewrite the following division problems in the standard form and divide.

1) 168 ÷ 6 = ______________

2) 477 ÷ 9 = ______________

3) 266 ÷ 7 = ______________

4) 480 ÷ 5 = ______________

5) 824 ÷ 8 = ______________

6) 864 ÷ 4 = ______________

7) 1,290 ÷ 10 = ______________

8) 1,771 ÷ 7 = ______________

9) 1,008 ÷ 9 = ______________

10) 3,069 ÷ 9 = ______________

11) 948 ÷ 12 = ______________

12) 1,472 ÷ 16 = ______________

13) 1,360 ÷ 16 = ______________

14) 2,160 ÷ 12 = ______________

15) 3,036 ÷ 6 = ______________

16) 8,844 ÷ 11 = ______________

17) 6,030 ÷ 3 = ______________

18) 5,400 ÷ 6 = ______________

19) 1,710 ÷ 6 = ______________

20) 1,820 ÷ 13 = ______________

21) 14,910 ÷ 21 = ______________

22) 15,625 ÷ 25 = ______________

23) 12,720 ÷ 12 = ______________

24) 13,797 ÷ 27 = ______________

25) 27,060 ÷ 60 = ______________

26) 21,350 ÷ 35 = ______________

27) 11,216 ÷ 16 = ______________

28) 12,030 ÷ 30 = ______________

29) 117,030 ÷ 30 = ______________

30) 43,026 ÷ 71 = ______________

31) 7,680 ÷ 64 = ______________

32) 162,054 ÷ 54 = ______________

▶ Solve these word problems with division.

33) The Jiffy Messenger Service traveled a total of 2,954 miles in one 7-day week. How many miles did they average daily? ________

34) Marvin collected 1,170 bottle tops over a 45-day period. How many bottle tops did he average per day? ________

Example:

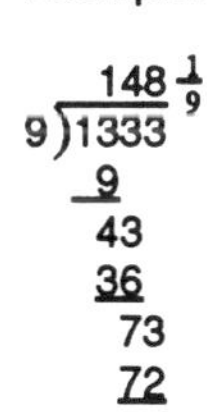

Name ____________________

Date ______________

MORE DIVISION OF WHOLE NUMBERS

▶ Rewrite the following division problems in the standard form and divide. Express the remainders in fractional form.

1) 1333 ÷ 9 = ____________

2) 898 ÷ 6 = ____________

3) 415 ÷ 6 = ____________

4) 2,115 ÷ 11 = ____________

5) 749 ÷ 6 = ____________

6) 1,218 ÷ 12 = ____________

7) 863 ÷ 7 = ____________

8) 3,017 ÷ 15 = ____________

9) 6,915 ÷ 4 = ____________

10) 812 ÷ 82 = ____________

11) 1,367 ÷ 17 = ____________

12) 3,575 ÷ 28 = ____________

13) 1,992 ÷ 10 = ____________

14) 2,115 ÷ 63 = ____________

15) 8,175 ÷ 35 = ____________

16) 7,167 ÷ 20 = ____________

17) 9,063 ÷ 75 = ____________

18) 10,613 ÷ 53 = ____________

19) 8,891 ÷ 22 = ____________

20) 7,776 ÷ 78 = ____________

21) 3,820 ÷ 45 = ____________

22) 29,666 ÷ 30 = ____________

23) 6,770 ÷ 65 = ____________

24) 41,080 ÷ 80 = ____________

25) 12,161 ÷ 11 = ____________

26) 58,775 ÷ 40 = ____________

27) 23,815 ÷ 15 = ____________

28) 91,090 ÷ 90 = ____________

29) 42,661 ÷ 36 = ____________

30) 39,042 ÷ 13 = ____________

▶ Solve these word problems with division. Express remainders in fractional form.

31) Lesley's father owned his car for 9 years, driving a total of 136,910 miles. Find the average number of miles driven per year. ________

32) Dottie drove her car 816 miles using 22 gallons of gas. Compute Dottie's gas mileage by dividing the number of miles driven by the number of gallons used. ________

Example:

$$\begin{array}{r} 48 \\ 10\overline{)480} \\ \underline{40} \\ 80 \\ \underline{80} \\ 0 \end{array}$$

or move decimal point one space to the left

48.0

Name ____________________

Date ____________

DIVIDING NUMBERS BY POWERS OF TEN

▶ Divide. Write the answers.

1) 480 ÷ 10 = ____________

2) 65,000 ÷ 100 = ____________

3) 2,000 ÷ 100 = ____________

4) 4,630 ÷ 10 = ____________

5) 9,600 ÷ 100 = ____________

6) 140,000 ÷ 1,000 = ____________

7) 191,000 ÷ 10 = ____________

8) 920,000 ÷ 100 = ____________

9) 62,000 ÷ 100 = ____________

10) 35,600 ÷ 100 = ____________

11) 385,000 ÷ 100 = ____________

12) 191,100,000 ÷ 1,000 = ____________

13) 25,000,000 ÷ 1,000 = ____________

14) 4,000,000 ÷ 1,000 = ____________

15) 806,000 ÷ 10 = ____________

16) 962,000 ÷ 100 = ____________

17) 305,000 ÷ 100 = ____________

18) 1,800,000 ÷ 1,000 = ____________

19) 600,000 ÷ 1,000 = ____________

20) 581,000 ÷ 10 = ____________

21) 720,600 ÷ 100 = ____________

22) 451,000 ÷ 1,000 = ____________

23) 390,000 ÷ 10 = ____________

24) 680,000 ÷ 100 = ____________

25) 4,060,300 ÷ 10 = ____________

26) 19,600 ÷ 10 = ____________

27) 9,603,000 ÷ 1,000 = ____________

28) 5,000,000 ÷ 100 = ____________

29) 7,000,000 ÷ 10 = ____________

30) 8,000,000 ÷ 100 = ____________

31) 123,000 ÷ 1,000 = ____________

32) 96,000,000 ÷ 1,000 = ____________

33) 43,000 ÷ 1,000 = ____________

34) 43,070,600 ÷ 10 = ____________

35) 8,000,000 ÷ 10,000 = ____________

36) 24,000,000,000 ÷ 10,000 = ____________

37) 902,000 ÷ 100 = ____________

38) 10,000,000 ÷ 10,000 = ____________

39) 304,000,000 ÷ 100 = ____________

40) 201,111,000 ÷ 10 = ____________

41) 76,000,000 ÷ 1,000 = ____________

42) 50,000 ÷ 10,000 = ____________

43) 240,000 ÷ 10,000 = ____________

44) 41,000,000 ÷ 100,000 = ____________

45) 101,000,000 ÷ 10 = ____________

46) 56,000,000 ÷ 10,000 = ____________

47) 28,000 ÷ 1,000 = ____________

48) 9,000,000,000 ÷ 100 = ____________

Name ____________________

Date ____________

BASIC OPERATIONS WITH WHOLE NUMBERS

▶ Add.

Example: Add
```
  22
 354
 356
 765
+ 87
1562
```

1) 354 + 356 + 765 + 87 = ________

2) 5,067 + 23 + 505 + 40 = ________

3) 1,012 + 705 + 4,033 = ________

4) 5,122 + 567 + 504 + 3,402 = ________

5) 243 + 2,023 + 4,050 + 5,670 = ________

6) 30,304 + 4,030 + 20,300 + 1,102 = ________

7) 203,340 + 94,059 + 304,450 = ________

8) 2,000 + 90,089 + 50,481 = ________

9) 10,223 + 4,055 + 506 + 8,690 = ________

10) 60,551 + 50,667 + 8,048,604 = ________

▶ Subtract.

Subtract
```
 112914
  2304
- 567
 1737
```

1) 2,304 − 567 = ________

2) 304,119 − 4,053 = ________

3) 30,400 − 19,234 = ________

4) 102,556 − 9,806 = ________

5) 203,445 − 12,789 = ________

6) 134,505 − 5,968 = ________

7) 900,800 − 203,788 = ________

8) 6,578,009 − 456,801 = ________

9) 340,599 − 9,875 = ________

10) 103,005 − 56,097 = ________

▶ Multiply.

Multiply
```
   1
  23
 x44
 092
 92
1012
```

1) 23 × 44 = ________

2) 304 × 32 = ________

3) 579 × 23 = ________

4) 3,011 × 44 = ________

5) 304 × 581 = ________

6) 4,503 × 23 = ________

7) 4,053 × 206 = ________

8) 5,098 × 2,304 = ________

9) 40,577 × 3,092 = ________

10) 9,056 × 3,419 = ________

▶ Divide. Write the remainders in the fractional form.

Divide
```
   1145
3)3435
  3
   4
   3
   13
   12
    15
    15
```

1) 3,435 ÷ 3 = ________

2) 2,034 ÷ 9 = ________

3) 49,571 ÷ 9 = ________

4) 30,455 ÷ 5 = ________

5) 30,987 ÷ 23 = ________

6) 46,570 ÷ 45 = ________

7) 30,575 ÷ 25 = ________

8) 192,304 ÷ 56 = ________

9) 340,987 ÷ 27 = ________

10) 987,230 ÷ 201 = ________

Example:

```
 25            63.16 = 63.2
 73         6)379.00
 80           36
 73            19
 33            18
+95             10
379              6
                 40
                 36
```

Name ______________________

Date ______________

AVERAGES

► Compute the averages for the sets of numbers. Round to the nearest tenth.

1) 25, 73, 80, 73, 33, 95 ____

2) 83, 78, 53, 92, 67, 27 ____

3) 20, 28, 19, 31, 22, 18, 17, 30 ____

4) 78, 98, 77, 67, 75, 90, 80, 90, 80, 75, 70 ____

5) 30, 31, 37, 33, 38, 35, 32, 39, 34, 36 ____

6) 44, 46, 64, 66, 62, 69, 41, 40 ____

7) 103, 110, 152, 173, 177, 100 150, 175, 152 ____

8) 205, 273, 198, 350, 220, 180, 280, 220 ____

9) 58, 74, 47, 83, 65, 36, 45, 46 70, 53, 55, 38 ____

10) 163, 219, 300, 512, 375, 602, 735, 638, 881 ____

11) 3004, 3210, 3387, 3652, 3470, 3521, 3980, 3922 ____

12) 5738, 5755, 5746, 5789, 5736, 5725, 5756, 5731 ____

13) 230, 310, 222, 725, 600, 390, 512, 525, 510, 400, 500, 683 ____

14) 3021, 5361, 2630, 6110, 4002, 3006, 4102, 4120, 4972, 3500, 2310 ____

15) 4589, 4530, 4520, 4500, 4530, 4528, 4501, 4554 ____

16) 7800, 7853, 7835, 7850, 7812, 7856, 7851, 7820 ____

17) 35, 78, 95, 83, 62, 89, 35, 60, 40, 66, 10, 31, 62, 89, 95 ____

18) 53, 62, 80, 22, 70, 35, 67, 89, 53, 60, 53, 67, 22, 70, 82, 10 ____

19) 304, 300, 384, 346, 320, 390, 355, 360, 340, 370, 300, 312 ____

20) 1239, 1264, 1220, 1250, 1235, 1260, 1285, 1240, 1200, 1290, 1206 ____

► Solve these word problems by computing the average. Round the answers to the nearest tenth.

21) Dan worked as a part-time car mechanic and worked 26 hours for his first week. What was his average number of hours worked per day if he worked four days? ______

22) The Camp Fire Girls sold 523 tickets to their annual jamboree. If there were 13 girls selling tickets, find the average number of tickets sold by each girl. ______

Example:

$2^3 = 2 \times 2 \times 2 = 8$

Name ______________________

Date ______________

EXPONENTS

► Express the following without exponents.

1) $2^3 =$ ______
2) $4^2 =$ ______
3) $5^3 =$ ______
4) $4^3 =$ ______
5) $6^2 =$ ______
6) $10^2 =$ ______
7) $8^2 =$ ______
8) $2^5 =$ ______
9) $9^3 =$ ______
10) $5^2 =$ ______
11) $4^4 =$ ______
12) $2^4 =$ ______
13) $8^3 =$ ______
14) $9^2 =$ ______
15) $7^2 =$ ______
16) $10^5 =$ ______
17) $3^4 =$ ______
18) $6^3 =$ ______
19) $7^3 =$ ______
20) $11^2 =$ ______
21) $20^3 =$ ______
22) $3^2 =$ ______
23) $5^4 =$ ______
24) $12^2 =$ ______
25) $10^3 =$ ______
26) $3^5 =$ ______
27) $22^3 =$ ______
28) $17^2 =$ ______
29) $15^2 =$ ______
30) $10^4 =$ ______
31) $12^3 =$ ______
32) $13^2 =$ ______
33) $20^2 =$ ______
34) $25^2 =$ ______
35) $15^3 =$ ______
36) $22^2 =$ ______
37) $11^3 =$ ______
38) $6^4 =$ ______
39) $2^6 =$ ______
40) $18^2 =$ ______
41) $4^5 =$ ______
42) 21^2 ______
43) $3^3 =$ ______
44) $10^6 =$ ______
45) $23^2 =$ ______
46) $14^2 =$ ______
47) $50^2 =$ ______
48) $100^2 =$ ______
49) $19^2 =$ ______
50) $33^2 =$ ______
51) $13^3 =$ ______
52) $16^2 =$ ______
53) $25^3 =$ ______
54) $17^3 =$ ______
55) $14^3 =$ ______
56) $5^5 =$ ______
57) $10^7 =$ ______
58) $11^4 =$ ______
59) $12^4 =$ ______
60) $10^9 =$ ______

Example:
$2 + 4 \times 2 =$
$2 + 8 = 10$

Name ____________________

Date ______________

ORDER OF OPERATIONS

▶ Find the answers. Perform the operations in the correct order.

1) $2 + 4 \times 2 =$ ________

2) $3 \times 4 + 6 - 4 =$ ________

3) $4 \times 8 + 16 \div 2 =$ ________

4) $5 \times 2 - 6 \div 2 =$ ________

5) $4^2 \times 2 + 5 - 32 =$ ________

6) $3 \times 2 \times 2^3 - 4^2 =$ ________

7) $7 + 6 \times 2 - 2 + 2^3 =$ ________

8) $18 - 2 \times 4^2 \div 4 =$ ________

9) $10 + 8 \times 6 \div 12 - 2 =$ ________

10) $13 - 4 \times 5 \div 2 + 10 =$ ________

11) $15 \times 3 - 5^2 + 10 =$ ________

12) $7^2 + 2^4 - 2^3 =$ ________

13) $4 + 17 - 3 \times 7 =$ ________

14) $12^2 - 10^2 + 5 \times 2 =$ ________

15) $10^2 - 2 \times 4 + 3^2 =$ ________

16) $11^2 + 23 - 2^3 + 9 =$ ________

17) $20 - 12 \div 6 \times 3 =$ ________

18) $4^3 - 3 \times 12 \div 6 =$ ________

19) $15 + 4 - 11 + 2^5 =$ ________

20) $12 \times 2 \div 3 \times 2 + 3 =$ ________

21) $8 \times 6 \div 4 - 12 \div 6 =$ ________

22) $2^3 \times 3 \div 6 + 12 - 3 =$ ________

23) $45 \div 15 + 10 - 2^3 =$ ________

24) $15 \div 3 - 5 + 10^2 =$ ________

25) $12^2 \div 6 - 20 + 7 =$ ________

26) $8^2 + 9 \times 3 - 10 =$ ________

27) $18 \div 3^2 - 2 + 5^2 =$ ________

28) $5^3 \div 5 - 10 + 2^3 =$ ________

29) $81 \div 3^2 - 6 + 12 \div 2 =$ ________

30) $3 \times 6 \div 2 - 5 + 7 =$ ________

31) $10^2 \div 5^2 + 5 \times 6 \div 2 =$ ________

32) $25 \div 5^2 \times 5 + 5 - 10 =$ ________

33) $100 \div 10 \times 2^2 + 8 =$ ________

34) $18 - 9 \times 2 \div 3 + 3^2 =$ ________

35) $50 - 40 + 4 \times 7 =$ ________

36) $28 \div 4 \times 6 - 20 =$ ________

37) $3^3 \times 2^2 + 20 \div 2 =$ ________

38) $6^2 \times 3 \div 2 - 4^2 =$ ________

39) $14 \div 2 \times 3 - 21 =$ ________

40) $16 \times 2 \div 4 - 2 + 10 =$ ________

Example:

$\frac{2}{3}$ and $\frac{4}{5}$: $2 \times 5 = 10$, $3 \times 4 = 12$

$\overset{10}{\frac{2}{3}} < \overset{12}{\frac{4}{5}}$

Name ______________________

Date ______________

COMPARING FRACTIONS

► Compare the fractions in each pair. Use < or > for each pair.

1) $\frac{2}{3}$ $\frac{4}{5}$	2) $\frac{5}{6}$ $\frac{7}{8}$	3) $\frac{5}{7}$ $\frac{7}{8}$	4) $\frac{2}{5}$ $\frac{4}{1}$
5) $\frac{3}{13}$ $\frac{6}{20}$	6) $\frac{3}{8}$ $\frac{9}{20}$	7) $\frac{6}{11}$ $\frac{5}{9}$	8) $\frac{5}{8}$ $\frac{10}{17}$
9) $\frac{1}{2}$ $\frac{5}{11}$	10) $\frac{2}{13}$ $\frac{4}{15}$	11) $\frac{15}{16}$ $\frac{16}{17}$	12) $\frac{2}{13}$ $\frac{3}{14}$
13) $\frac{1}{8}$ $\frac{2}{13}$	14) $\frac{5}{21}$ $\frac{18}{31}$	15) $\frac{2}{15}$ $\frac{6}{21}$	16) $\frac{7}{13}$ $\frac{6}{32}$
17) $\frac{1}{5}$ $\frac{1}{9}$	18) $\frac{2}{21}$ $\frac{3}{31}$	19) $\frac{1}{3}$ $\frac{2}{7}$	20) $\frac{6}{7}$ $\frac{8}{1}$
21) $\frac{3}{5}$ $\frac{1}{2}$	22) $\frac{2}{8}$ $\frac{4}{17}$	23) $\frac{6}{10}$ $\frac{1}{5}$	24) $\frac{5}{10}$ $\frac{10}{21}$
25) $\frac{4}{7}$ $\frac{5}{11}$	26) $\frac{3}{5}$ $\frac{6}{11}$	27) $\frac{4}{5}$ $\frac{2}{20}$	28) $\frac{6}{11}$ $\frac{5}{22}$
29) $\frac{1}{2}$ $\frac{2}{8}$	30) $\frac{4}{11}$ $\frac{8}{21}$	31) $\frac{9}{10}$ $\frac{4}{5}$	32) $\frac{7}{12}$ $\frac{7}{10}$
33) $\frac{6}{9}$ $\frac{12}{17}$	34) $\frac{4}{12}$ $\frac{2}{7}$	35) $\frac{3}{13}$ $\frac{1}{6}$	36) $\frac{8}{15}$ $\frac{15}{16}$
37) $\frac{3}{8}$ $\frac{5}{13}$	38) $\frac{10}{12}$ $\frac{14}{16}$	39) $\frac{2}{6}$ $\frac{4}{14}$	40) $\frac{1}{12}$ $\frac{4}{24}$
41) $\frac{6}{11}$ $\frac{5}{13}$	42) $\frac{2}{4}$ $\frac{10}{22}$	43) $\frac{8}{10}$ $\frac{4}{40}$	44) $\frac{6}{7}$ $\frac{18}{20}$
45) $\frac{4}{6}$ $\frac{8}{10}$	46) $\frac{6}{10}$ $\frac{2}{4}$	47) $\frac{11}{23}$ $\frac{22}{31}$	48) $\frac{9}{15}$ $\frac{18}{31}$
49) $\frac{10}{11}$ $\frac{20}{23}$	50) $\frac{4}{8}$ $\frac{5}{11}$	51) $\frac{10}{23}$ $\frac{15}{32}$	52) $\frac{11}{13}$ $\frac{5}{17}$
53) $\frac{9}{10}$ $\frac{13}{14}$	54) $\frac{7}{11}$ $\frac{8}{12}$	55) $\frac{3}{7}$ $\frac{5}{9}$	56) $\frac{5}{6}$ $\frac{16}{17}$
57) $\frac{2}{11}$ $\frac{2}{3}$	58) $\frac{1}{8}$ $\frac{5}{45}$	59) $\frac{13}{20}$ $\frac{1}{5}$	60) $\frac{7}{10}$ $\frac{6}{15}$

Name ______________________________

Date ________________

Example:

$\frac{2}{5} = \frac{}{25}$ Divide 25 by 5. $25 \div 5 = 5$

$\frac{2}{5} \times \frac{5}{5} = \frac{10}{25}$

$\frac{2}{5} = \frac{10}{25}$

WORKING WITH FRACTIONS

► Express these fractions in higher terms.

1) $\frac{2}{5} = \frac{}{25}$ 2) $\frac{1}{3} = \frac{}{18}$ 3) $\frac{5}{6} = \frac{}{24}$ 4) $\frac{7}{8} = \frac{}{32}$

5) $\frac{7}{8} = \frac{}{56}$ 6) $\frac{3}{7} = \frac{}{21}$ 7) $\frac{2}{9} = \frac{}{36}$ 8) $\frac{1}{5} = \frac{}{30}$

9) $\frac{1}{4} = \frac{}{20}$ 10) $\frac{5}{11} = \frac{}{121}$ 11) $\frac{4}{9} = \frac{}{72}$ 12) $\frac{3}{11} = \frac{}{44}$

13) $\frac{2}{3} = \frac{}{18}$ 14) $\frac{7}{10} = \frac{}{80}$ 15) $\frac{3}{4} = \frac{}{16}$ 16) $\frac{6}{7} = \frac{}{42}$

17) $\frac{7}{13} = \frac{}{39}$ 18) $\frac{5}{8} = \frac{}{64}$ 19) $\frac{2}{13} = \frac{}{52}$ 20) $\frac{1}{9} = \frac{}{99}$

21) $\frac{11}{12} = \frac{}{60}$ 22) $\frac{8}{15} = \frac{}{45}$ 23) $\frac{6}{19} = \frac{}{76}$ 24) $\frac{4}{7} = \frac{}{63}$

25) $\frac{13}{15} = \frac{}{105}$ 26) $\frac{8}{17} = \frac{}{34}$ 27) $\frac{10}{19} = \frac{}{38}$ 28) $\frac{1}{13} = \frac{}{65}$

29) $\frac{10}{11} = \frac{}{55}$ 30) $\frac{15}{16} = \frac{}{64}$ 31) $\frac{5}{14} = \frac{}{56}$ 32) $\frac{2}{23} = \frac{}{92}$

33) $\frac{5}{7} = \frac{}{210}$ 34) $\frac{4}{13} = \frac{}{39}$ 35) $\frac{5}{10} = \frac{}{1000}$ 36) $\frac{9}{17} = \frac{}{51}$

37) $\frac{4}{11} = \frac{}{99}$ 38) $\frac{23}{30} = \frac{}{90}$ 39) $\frac{3}{7} = \frac{}{84}$ 40) $\frac{3}{16} = \frac{}{80}$

41) $\frac{12}{15} = \frac{}{60}$ 42) $\frac{5}{6} = \frac{}{54}$ 43) $\frac{8}{14} = \frac{}{42}$ 44) $\frac{18}{23} = \frac{}{92}$

45) $\frac{9}{22} = \frac{}{220}$ 46) $\frac{5}{8} = \frac{}{96}$ 47) $\frac{5}{21} = \frac{}{105}$ 48) $\frac{5}{32} = \frac{}{160}$

49) $\frac{6}{38} = \frac{}{380}$ 50) $\frac{35}{70} = \frac{}{280}$ 51) $\frac{6}{30} = \frac{}{390}$ 52) $\frac{3}{13} = \frac{}{65}$

53) $\frac{5}{11} = \frac{}{121}$ 54) $\frac{45}{60} = \frac{}{480}$ 55) $\frac{15}{18} = \frac{}{72}$ 56) $\frac{8}{52} = \frac{}{156}$

57) $\frac{6}{16} = \frac{}{80}$ 58) $\frac{7}{12} = \frac{}{156}$ 59) $\frac{9}{29} = \frac{}{145}$ 60) $\frac{5}{21} = \frac{}{126}$

61) $\frac{5}{11} = \frac{}{99}$ 62) $\frac{3}{52} = \frac{}{208}$ 63) $\frac{2}{3} = \frac{}{369}$ 64) $\frac{2}{40} = \frac{}{120}$

65) $\frac{9}{17} = \frac{}{34}$ 66) $\frac{11}{12} = \frac{}{120}$ 67) $\frac{6}{21} = \frac{}{84}$ 68) $\frac{17}{45} = \frac{}{360}$

69) $\frac{9}{13} = \frac{}{39}$ 70) $\frac{9}{15} = \frac{}{75}$ 71) $\frac{5}{30} = \frac{}{270}$ 72) $\frac{3}{11} = \frac{}{121}$

Name ____________________

Date ____________

Example:
$\frac{10}{14} = \frac{10 \div 2}{14 \div 2} = \frac{5}{7}$

RENAME THE FRACTIONS

► Express these fractions and mixed numbers in their lowest terms.

1) $\frac{10}{14} =$	2) $\frac{9}{27} =$	3) $\frac{18}{22} =$	4) $13\frac{10}{15} =$
5) $2\frac{12}{14} =$	6) $\frac{10}{20} =$	7) $\frac{13}{39} =$	8) $\frac{28}{40} =$
9) $\frac{10}{50} =$	10) $6\frac{5}{40} =$	11) $13\frac{70}{80} =$	12) $27\frac{52}{104} =$
13) $\frac{30}{33} =$	14) $9\frac{24}{46} =$	15) $\frac{9}{21} =$	16) $\frac{150}{200} =$
17) $20\frac{18}{57} =$	18) $\frac{46}{52} =$	19) $\frac{28}{120} =$	20) $\frac{102}{128} =$
21) $\frac{178}{220} =$	22) $8\frac{23}{92} =$	23) $51\frac{66}{121} =$	24) $9\frac{15}{33} =$
25) $\frac{44}{108} =$	26) $9\frac{31}{62} =$	27) $6\frac{7}{63} =$	28) $\frac{16}{64} =$
29) $\frac{88}{121} =$	30) $\frac{58}{64} =$	31) $11\frac{62}{92} =$	32) $18\frac{35}{225} =$
33) $\frac{28}{54} =$	34) $\frac{20}{55} =$	35) $\frac{36}{38} =$	36) $\frac{42}{150} =$
37) $\frac{20}{25} =$	38) $\frac{11}{88} =$	39) $\frac{33}{39} =$	40) $\frac{88}{112} =$
41) $\frac{38}{56} =$	42) $\frac{15}{54} =$	43) $\frac{56}{108} =$	44) $\frac{38}{106} =$
45) $\frac{25}{155} =$	46) $\frac{18}{450} =$	47) $\frac{210}{280} =$	48) $\frac{64}{160} =$
49) $\frac{30}{72} =$	50) $\frac{450}{480} =$	51) $\frac{22}{42} =$	52) $\frac{140}{280} =$
53) $\frac{150}{300} =$	54) $\frac{84}{96} =$	55) $\frac{50}{68} =$	56) $\frac{78}{104} =$
57) $\frac{41}{205} =$	58) $\frac{69}{207} =$	59) $\frac{42}{122} =$	60) $\frac{55}{300} =$
61) $\frac{18}{39} =$	62) $6\frac{50}{400} =$	63) $25\frac{148}{256} =$	64) $35\frac{110}{300} =$
65) $\frac{45}{78} =$	66) $18\frac{100}{150} =$	67) $2\frac{44}{56} =$	68) $2\frac{90}{99} =$
69) $\frac{380}{560} =$	70) $32\frac{32}{80} =$	71) $7\frac{35}{210} =$	72) $6\frac{80}{220} =$

Example:

$1\frac{2}{3} = \frac{5}{3}$

Multiply the whole number by the denominator.
Then, add the numerator.

Name ____________________

Date ____________

MIXED NUMBERS

► Rename these mixed numbers as improper fractions.

1) $1\frac{2}{3} =$	2) $1\frac{1}{2} =$	3) $2\frac{1}{5} =$	4) $1\frac{5}{6} =$
5) $4\frac{1}{5} =$	6) $3\frac{2}{5} =$	7) $1\frac{1}{6} =$	8) $9\frac{2}{7} =$
9) $4\frac{3}{4} =$	10) $2\frac{5}{11} =$	11) $1\frac{5}{9} =$	12) $13\frac{2}{7} =$
13) $20\frac{1}{2} =$	14) $6\frac{2}{9} =$	15) $3\frac{4}{7} =$	16) $2\frac{2}{3} =$
17) $15\frac{1}{3} =$	18) $6\frac{4}{11} =$	19) $15\frac{3}{4} =$	20) $6\frac{2}{5} =$
21) $7\frac{1}{5} =$	22) $33\frac{2}{3} =$	23) $17\frac{1}{2} =$	24) $2\frac{2}{17} =$
25) $9\frac{5}{11} =$	26) $8\frac{5}{13} =$	27) $18\frac{3}{5} =$	28) $3\frac{2}{19} =$
29) $9\frac{10}{11} =$	30) $2\frac{13}{14} =$	31) $3\frac{16}{17} =$	32) $4\frac{2}{19} =$
33) $5\frac{7}{9} =$	34) $5\frac{2}{12} =$	35) $6\frac{3}{13} =$	36) $4\frac{1}{13} =$
37) $7\frac{10}{11} =$	38) $8\frac{13}{15} =$	39) $1\frac{7}{22} =$	40) $3\frac{5}{11} =$
41) $4\frac{5}{20} =$	42) $2\frac{3}{22} =$	43) $5\frac{11}{16} =$	44) $21\frac{2}{61} =$
45) $5\frac{13}{20} =$	46) $7\frac{20}{23} =$	47) $5\frac{5}{60} =$	48) $1\frac{2}{13} =$
49) $3\frac{5}{11} =$	50) $7\frac{2}{25} =$	51) $4\frac{3}{29} =$	52) $4\frac{2}{32} =$
53) $1\frac{7}{18} =$	54) $11\frac{5}{11} =$	55) $20\frac{1}{16} =$	56) $10\frac{2}{17} =$
57) $11\frac{5}{21} =$	58) $5\frac{11}{12} =$	59) $7\frac{2}{30} =$	60) $2\frac{6}{23} =$
61) $2\frac{7}{18} =$	62) $2\frac{6}{11} =$	63) $8\frac{5}{13} =$	64) $28\frac{7}{12} =$
65) $9\frac{1}{76} =$	66) $3\frac{2}{19} =$	67) $8\frac{2}{17} =$	68) $12\frac{5}{82} =$

Example:

Rename $\frac{13}{5}$

$$5\overline{)13}\ = 2\frac{3}{5}$$
$$\underline{10}$$
$$3$$

Name ____________________

Date ____________

IMPROPER FRACTIONS TO MIXED NUMBERS

► Rename these improper fractions as mixed numbers.

1) $\frac{13}{5} =$	2) $\frac{16}{7} =$	3) $\frac{28}{7} =$	4) $\frac{15}{2} =$
5) $\frac{33}{4} =$	6) $\frac{18}{5} =$	7) $\frac{25}{4} =$	8) $\frac{62}{6} =$
9) $\frac{30}{7} =$	10) $\frac{35}{8} =$	11) $\frac{18}{8} =$	12) $\frac{30}{4} =$
13) $\frac{75}{25} =$	14) $\frac{72}{10} =$	15) $\frac{26}{3} =$	16) $\frac{72}{19} =$
17) $\frac{60}{9} =$	18) $\frac{120}{11} =$	19) $\frac{36}{17} =$	20) $\frac{57}{9} =$
21) $\frac{39}{16} =$	22) $\frac{37}{5} =$	23) $\frac{17}{2} =$	24) $\frac{60}{28} =$
25) $\frac{33}{10} =$	26) $\frac{135}{9} =$	27) $\frac{200}{120} =$	28) $\frac{25}{21} =$
29) $\frac{63}{12} =$	30) $\frac{130}{12} =$	31) $\frac{38}{6} =$	32) $\frac{17}{5} =$
33) $\frac{18}{11} =$	34) $\frac{23}{2} =$	35) $\frac{54}{11} =$	36) $\frac{92}{5} =$
37) $\frac{36}{13} =$	38) $\frac{39}{5} =$	39) $\frac{51}{11} =$	40) $\frac{46}{14} =$
41) $\frac{22}{5} =$	42) $\frac{87}{12} =$	43) $\frac{57}{12} =$	44) $\frac{62}{11} =$
45) $\frac{73}{10} =$	46) $\frac{91}{8} =$	47) $\frac{76}{7} =$	48) $\frac{89}{11} =$
49) $\frac{48}{23} =$	50) $\frac{98}{46} =$	51) $\frac{88}{23} =$	52) $\frac{105}{15} =$
53) $\frac{28}{6} =$	54) $\frac{100}{7} =$	55) $\frac{49}{3} =$	56) $\frac{65}{2} =$
57) $\frac{85}{17} =$	58) $\frac{59}{9} =$	59) $\frac{99}{10} =$	60) $\frac{46}{3} =$
61) $\frac{98}{5} =$	62) $\frac{47}{2} =$	63) $\frac{77}{12} =$	64) $\frac{98}{8} =$
65) $\frac{22}{62} =$	66) $\frac{68}{45} =$	67) $\frac{100}{98} =$	68) $\frac{210}{50} =$

Example:
$\frac{3}{5} \times \frac{7}{8} = \frac{3 \times 7}{5 \times 8} = \frac{21}{40}$

Name ______________________________

Date ________________

MULTIPLYING FRACTIONS

▶ Multiply these fractions. Simplify the answers to the lowest terms.

1) $\frac{3}{5} \times \frac{7}{8} =$

2) $\frac{4}{5} \times \frac{10}{11} =$

3) $\frac{6}{9} \times \frac{1}{3} =$

4) $\frac{2}{5} \times \frac{4}{10} =$

5) $8 \times \frac{6}{7} =$

6) $\frac{5}{7} \times \frac{3}{8} =$

7) $\frac{7}{9} \times 1\frac{1}{2} =$

8) $\frac{2}{9} \times 1\frac{1}{8} =$

9) $\frac{3}{11} \times 3\frac{2}{3} =$

10) $\frac{7}{13} \times \frac{13}{5} =$

11) $\frac{8}{11} \times \frac{1}{5} =$

12) $2\frac{1}{2} \times \frac{1}{2} =$

13) $\frac{2}{5} \times \frac{2}{17} =$

14) $\frac{5}{8} \times \frac{3}{8} =$

15) $\frac{4}{6} \times \frac{2}{3} =$

16) $\frac{1}{3} \times \frac{6}{10} =$

17) $\frac{1}{2} \times \frac{3}{5} =$

18) $\frac{2}{9} \times \frac{5}{11} =$

19) $\frac{35}{38} \times \frac{4}{5} =$

20) $\frac{6}{11} \times 2\frac{5}{6} =$

21) $\frac{7}{12} \times \frac{3}{7} =$

22) $\frac{1}{6} \times \frac{2}{5} =$

23) $\frac{2}{5} \times 2\frac{2}{5} =$

24) $6\frac{2}{7} \times \frac{2}{3} =$

25) $1\frac{3}{7} \times \frac{5}{8} =$

26) $\frac{6}{13} \times 2\frac{1}{6} =$

27) $3\frac{1}{5} \times \frac{3}{11} =$

28) $4\frac{1}{5} \times 7 =$

29) $5\frac{2}{5} \times 10 =$

30) $2\frac{3}{11} \times 22 =$

31) $5\frac{2}{13} \times \frac{1}{6} =$

32) $3\frac{4}{5} \times 65 =$

33) $5\frac{2}{13} \times \frac{1}{6} =$

34) $2\frac{1}{5} \times 5\frac{1}{2} =$

35) $7\frac{2}{5} \times 5\frac{2}{7} =$

36) $8\frac{2}{13} \times 13\frac{8}{9} =$

37) $2\frac{1}{20} \times 20\frac{5}{10} =$

38) $5\frac{6}{13} \times 1\frac{1}{2} =$

39) $7\frac{1}{5} \times \frac{65}{72} =$

40) $5\frac{2}{7} \times 1\frac{1}{2} =$

41) $7\frac{5}{8} \times 2\frac{3}{4} =$

42) $4\frac{6}{7} \times \frac{42}{68} =$

43) $5\frac{2}{12} \times 1\frac{1}{2} =$

44) $5\frac{4}{11} \times 1\frac{1}{3} =$

45) $3\frac{4}{5} \times 1\frac{1}{2} =$

46) $8\frac{8}{9} \times \frac{18}{20} =$

47) $3\frac{1}{5} \times 5\frac{1}{3} =$

48) $20\frac{1}{5} \times \frac{2}{3} =$

49) $3\frac{1}{3} \times \frac{1}{5} =$

50) $2\frac{3}{7} \times \frac{14}{17} =$

51) $3\frac{5}{6} \times \frac{12}{46} =$

52) $5\frac{2}{7} \times 5 =$

53) $7\frac{2}{5} \times 1\frac{1}{5} =$

54) $3\frac{2}{3} \times 1\frac{1}{2} =$

Example:

$\frac{3}{5} \div \frac{2}{5} = \frac{3}{5} \times \frac{5}{2}$

$= \frac{3}{\not 5_1} \times \frac{\not 5^1}{2}$

$= \frac{3 \times 1}{1 \times 2}$

$= \frac{3}{2}$

$= 1\frac{1}{2}$

Name ____________________

Date ______________

DIVIDING FRACTIONS

▶ Divide these fractions. Simplify the answers to the lowest terms.

1) $\frac{3}{5} \div \frac{2}{5} =$	2) $\frac{5}{7} \div \frac{1}{2} =$	3) $\frac{6}{7} \div \frac{2}{9} =$
4) $\frac{5}{13} \div \frac{25}{26} =$	5) $\frac{15}{16} \div \frac{3}{8} =$	6) $1\frac{2}{3} \div \frac{3}{5} =$
7) $\frac{6}{13} \div \frac{2}{13} =$	8) $\frac{3}{10} \div 1\frac{4}{5} =$	9) $\frac{6}{11} \div 2\frac{1}{5} =$
10) $\frac{4}{5} \div \frac{24}{25} =$	11) $\frac{5}{7} \div \frac{14}{15} =$	12) $\frac{2}{13} \div \frac{6}{7} =$
13) $\frac{1}{2} \div \frac{1}{3} =$	14) $\frac{5}{2} \div 1\frac{1}{2} =$	15) $\frac{4}{3} \div 2\frac{1}{5} =$
16) $\frac{1}{5} \div \frac{3}{4} =$	17) $\frac{1}{12} \div \frac{2}{3} =$	18) $\frac{9}{10} \div \frac{2}{5} =$
19) $\frac{2}{5} \div \frac{1}{8} =$	20) $1\frac{2}{7} \div 2 =$	21) $5\frac{2}{8} \div \frac{1}{8} =$
22) $\frac{12}{17} \div \frac{15}{21} =$	23) $\frac{1}{3} \div 2\frac{1}{2} =$	24) $23\frac{1}{2} \div \frac{1}{2} =$
25) $2\frac{3}{4} \div \frac{11}{16} =$	26) $5\frac{3}{5} \div 2\frac{1}{10} =$	27) $2\frac{1}{5} \div \frac{1}{5} =$
28) $3\frac{2}{7} \div \frac{46}{21} =$	29) $1\frac{2}{5} \div 2\frac{2}{3} =$	30) $3\frac{2}{7} \div 1\frac{2}{5} =$
31) $1\frac{2}{8} \div 2\frac{1}{2} =$	32) $\frac{5}{16} \div 1\frac{1}{2} =$	33) $7\frac{1}{2} \div 22\frac{1}{2} =$
34) $5\frac{2}{3} \div 1\frac{5}{12} =$	35) $3\frac{5}{7} \div 5\frac{4}{7} =$	36) $\frac{3}{7} \div 1\frac{3}{11} =$
37) $2\frac{1}{6} \div 1\frac{1}{2} =$	38) $1\frac{2}{11} \div 1\frac{1}{22} =$	39) $3\frac{5}{13} \div \frac{22}{39} =$
40) $5 \div 1\frac{1}{6} =$	41) $11 \div 3\frac{4}{33} =$	42) $3\frac{2}{11} \div \frac{1}{6} =$
43) $4\frac{3}{11} \div 6 =$	44) $4\frac{1}{2} \div \frac{36}{38} =$	45) $3\frac{2}{7} \div 1\frac{1}{2} =$
46) $2\frac{5}{6} \div 17 =$	47) $1\frac{5}{9} \div 1\frac{2}{5} =$	48) $3\frac{1}{6} \div 1\frac{1}{6} =$
49) $2\frac{3}{5} \div \frac{1}{5} =$	50) $5\frac{3}{8} \div 1\frac{1}{16} =$	51) $1\frac{3}{11} \div 5 =$
52) $1\frac{1}{7} \div \frac{3}{7} =$	53) $8\frac{1}{3} \div 1\frac{1}{9} =$	54) $5\frac{2}{9} \div 2\frac{1}{2} =$

Example:

$$\begin{array}{r} \frac{5}{12} \\ + \frac{6}{12} \\ \hline \frac{11}{12} \end{array}$$

Name ____________________

Date ____________

ADDING FRACTIONS

► Add these fractions. Simplify the answers to the lowest terms.

1) $\frac{5}{12} + \frac{6}{12}$

2) $\frac{2}{9} + \frac{3}{9}$

3) $\frac{5}{13} + \frac{8}{13}$

4) $4 + 6\frac{2}{5}$

5) $\frac{3}{16} + \frac{2}{16}$

6) $\frac{2}{17} + \frac{5}{17}$

7) $\frac{7}{18} + \frac{2}{18}$

8) $7\frac{2}{13} + \frac{11}{13}$

9) $\frac{5}{15} + \frac{3}{15}$

10) $\frac{3}{20} + \frac{2}{20}$

11) $\frac{6}{20} + \frac{5}{20}$

12) $6\frac{2}{7} + 5\frac{3}{7}$

13) $\frac{7}{21} + \frac{7}{21}$

14) $2\frac{2}{18} + 3\frac{3}{18}$

15) $4\frac{6}{23} + 5\frac{17}{23}$

16) $2\frac{5}{8} + 3$

17) $\frac{8}{9} + \frac{5}{9}$

18) $6\frac{7}{10} + \frac{9}{10}$

19) $5\frac{5}{26} + \frac{24}{26}$

20) $6\frac{2}{15} + 4\frac{8}{15}$

21) $5\frac{2}{3} + 2\frac{1}{3}$

22) $12\frac{7}{11} + 3\frac{2}{11}$

23) $35\frac{5}{6} + 5$

24) $3\frac{5}{13} + 4\frac{2}{13}$

Example:
Find the least common multiple of the denominators.
Then, raise the fractions to higher terms.

$$\begin{array}{rcr} \frac{5}{6} & = & \frac{15}{18} \\ +\frac{2}{18} & = & +\frac{2}{18} \\ \hline & & \frac{17}{18} \end{array}$$

Name ______________________

Date ______________

FRACTIONS WITH UNLIKE DENOMINATORS

►Add these fractions. Simplify the answers.

1) $\frac{5}{6} + \frac{2}{18}$

2) $\frac{5}{9} + \frac{4}{18}$

3) $\frac{6}{25} + \frac{4}{5}$

4) $\frac{12}{35} + 7\frac{1}{210}$

5) $\frac{7}{18} + \frac{1}{3}$

6) $\frac{4}{21} + \frac{1}{7}$

7) $\frac{8}{35} + \frac{2}{7}$

8) $18\frac{1}{6} + 3\frac{1}{72}$

9) $5\frac{3}{7} + 2\frac{1}{14}$

10) $6\frac{3}{20} + 4\frac{2}{5}$

11) $8\frac{3}{20} + 5\frac{3}{4}$

12) $8\frac{3}{21} + 2\frac{1}{7}$

13) $10\frac{2}{17} + 3\frac{5}{34}$

14) $23\frac{1}{7} + 2\frac{5}{14}$

15) $37\frac{4}{8} + 5\frac{7}{24}$

16) $8\frac{1}{5} + 10\frac{1}{55}$

17) $25\frac{6}{35} + 4\frac{3}{70}$

18) $38\frac{1}{12} + 2\frac{4}{60}$

19) $3\frac{3}{5} + 2\frac{1}{20}$

20) $12\frac{5}{28} + 4\frac{5}{56}$

21) $8\frac{2}{7} + 7\frac{3}{56}$

22) $4\frac{5}{11} + 1\frac{5}{66}$

23) $6\frac{4}{9} + \frac{1}{3}$

24) $5\frac{5}{26} + 2\frac{1}{104}$

Example:

$$\frac{5}{9} = \frac{35}{63}$$
$$+\frac{3}{7} = +\frac{27}{63}$$
$$\frac{62}{63}$$

Name ____________________

Date ____________

MORE FRACTIONS WITH UNLIKE DENOMINATORS

► Add these fractions. Simplify the answers.

1) $\frac{5}{9} + \frac{3}{7}$

2) $\frac{3}{13} + \frac{1}{5}$

3) $\frac{6}{22} + \frac{4}{5}$

4) $3\frac{4}{7} + 7\frac{3}{8}$

5) $\frac{5}{11} + \frac{4}{7}$

6) $\frac{3}{7} + \frac{9}{9}$

7) $\frac{7}{11} + \frac{10}{12}$

8) $28\frac{5}{7} + 6\frac{4}{9}$

9) $2\frac{5}{7} + 5\frac{6}{8}$

10) $\frac{5}{12} + 6\frac{4}{10}$

11) $2\frac{7}{13} + 9\frac{3}{10}$

12) $4\frac{1}{12} + 6\frac{2}{7}$

13) $5\frac{4}{9} + 3\frac{2}{11}$

14) $6\frac{3}{13} + 4\frac{5}{6}$

15) $3\frac{5}{7} + 2\frac{9}{11}$

16) $35\frac{4}{19} + 2\frac{3}{10}$

17) $16\frac{5}{12} + 4\frac{3}{5}$

18) $9\frac{5}{11} + 6\frac{2}{5}$

19) $4\frac{7}{13} + 2\frac{3}{5}$

20) $12\frac{8}{15} + \frac{1}{4}$

21) $9\frac{11}{20} + 6\frac{4}{7}$

22) $13\frac{12}{15} + 5\frac{3}{4}$

23) $2\frac{1}{2} + 6\frac{3}{27}$

24) $14\frac{2}{21} + 5\frac{1}{2}$

Name ______________________

Date ______________

Example:

$$\begin{array}{r} 3\frac{2}{13} \\ +4\frac{1}{13} \\ \hline 7\frac{3}{13} \end{array}$$

ADDITION OF FRACTIONS

▶ Add these fractions. Simplify the answers to the lowest terms.

1) $\begin{array}{r} 3\frac{2}{13} \\ +\ 4\frac{1}{13} \\ \hline \end{array}$

2) $\begin{array}{r} 4\frac{2}{5} \\ +\ 3\frac{1}{5} \\ \hline \end{array}$

3) $\begin{array}{r} 6\frac{5}{21} \\ +\ 2\frac{1}{7} \\ \hline \end{array}$

4) $\begin{array}{r} 18\frac{5}{16} \\ +\ 3\frac{1}{8} \\ \hline \end{array}$

5) $\begin{array}{r} 7\frac{1}{4} \\ +\ \frac{1}{3} \\ \hline \end{array}$

6) $\begin{array}{r} 13\frac{5}{16} \\ +\ 2\frac{1}{4} \\ \hline \end{array}$

7) $\begin{array}{r} 81\frac{2}{15} \\ +\ 2\frac{2}{3} \\ \hline \end{array}$

8) $\begin{array}{r} 17 \\ +\ 2\frac{1}{5} \\ \hline \end{array}$

9) $\begin{array}{r} 10\frac{3}{7} \\ +\ 2\frac{4}{5} \\ \hline \end{array}$

10) $\begin{array}{r} 18\frac{2}{11} \\ +\ 4\frac{5}{6} \\ \hline \end{array}$

11) $\begin{array}{r} 29\frac{2}{5} \\ +\ 3\frac{5}{6} \\ \hline \end{array}$

12) $\begin{array}{r} 13\frac{3}{22} \\ +\ 5\frac{3}{44} \\ \hline \end{array}$

13) $\begin{array}{r} 23\frac{4}{7} \\ +\ 8 \\ \hline \end{array}$

14) $\begin{array}{r} 16\frac{3}{13} \\ +\ \frac{4}{39} \\ \hline \end{array}$

15) $\begin{array}{r} 11\frac{2}{9} \\ +\ 5\frac{3}{8} \\ \hline \end{array}$

16) $\begin{array}{r} 9\frac{1}{9} \\ +\ 5\frac{4}{27} \\ \hline \end{array}$

17) $\begin{array}{r} 1\frac{5}{11} \\ +\ 3\frac{4}{7} \\ \hline \end{array}$

18) $\begin{array}{r} 19\frac{9}{13} \\ +\ 2\frac{1}{3} \\ \hline \end{array}$

▶ Rewrite the fractions in the standard form and add. Simplify the answers to the lowest terms.

19) $1\frac{1}{2} + 2\frac{3}{11} =$

20) $2\frac{3}{5} + 4\frac{1}{6} =$

21) $7\frac{1}{7} + 5\frac{1}{8} =$

22) $6\frac{2}{9} + 1\frac{3}{10} =$

23) $10 + 2\frac{1}{6} =$

24) $5\frac{2}{12} + \frac{1}{10} =$

Example:

$$\begin{array}{r} \frac{7}{8} \\ -\ \frac{3}{8} \\ \hline \frac{4}{8} = \frac{1}{2} \end{array}$$

Name ____________________

Date ____________

FRACTIONS WITH LIKE DENOMINATORS

▶ Subtract these fractions and simplify your answers.

1) $\frac{7}{8} - \frac{3}{8}$

2) $\frac{13}{16} - \frac{5}{16}$

3) $\frac{10}{21} - \frac{7}{21}$

4) $13\frac{4}{29} - 3\frac{2}{29}$

5) $3\frac{7}{12} - 2\frac{2}{12}$

6) $5\frac{15}{18} - 2\frac{3}{18}$

7) $11\frac{4}{10} - \frac{1}{10}$

8) $8\frac{33}{56} - 2\frac{5}{56}$

9) $8\frac{13}{15} - 3\frac{4}{15}$

10) $11\frac{12}{41} - 5\frac{6}{41}$

11) $30\frac{18}{33} - 5\frac{7}{33}$

12) $33\frac{37}{45} - \frac{2}{45}$

13) $16\frac{11}{20} - \frac{6}{20}$

14) $8\frac{5}{28} - 5$

15) $31\frac{14}{27} - 4\frac{5}{27}$

16) $13\frac{17}{38} - \frac{7}{38}$

17) $7\frac{2}{33} - 5$

18) $13\frac{17}{18} - 4\frac{8}{18}$

19) $25\frac{3}{16} - 4\frac{1}{16}$

20) $25\frac{7}{33} - 4\frac{4}{33}$

21) $9\frac{40}{52} - 4\frac{14}{52}$

22) $23\frac{39}{40} - 4\frac{7}{40}$

23) $19\frac{26}{27} - 1\frac{7}{27}$

24) $22\frac{18}{35} - 5\frac{9}{35}$

Example:

$$\begin{array}{r} 15\frac{2}{3} = 15\frac{14}{21} \\ -\ 6\frac{2}{7} = \ \ 6\frac{6}{21} \\ \hline 9\frac{8}{21} \end{array}$$

Name ____________________________

Date ________________

SUBTRACTION OF FRACTIONS WITHOUT RENAMING

▶Subtract these fractions. Simplify the answers to the lowest terms.

1) $15\frac{2}{3} - 6\frac{2}{7}$

2) $22\frac{7}{8} - 3\frac{1}{24}$

3) $25\frac{12}{38} - 9\frac{3}{19}$

4) $8\frac{1}{2} - 5\frac{3}{9}$

5) $28\frac{12}{15} - 4\frac{2}{5}$

6) $26\frac{4}{9} - 4\frac{1}{8}$

7) $33\frac{4}{7} - 6\frac{1}{7}$

8) $42\frac{6}{7} - 5\frac{1}{6}$

9) $52\frac{3}{10} - 2\frac{5}{17}$

10) $14\frac{5}{11} - 7\frac{2}{7}$

11) $4\frac{11}{16} - 2\frac{3}{32}$

12) $2\frac{1}{3} - 1\frac{2}{15}$

13) $3\frac{15}{21} - 1\frac{1}{5}$

14) $7\frac{2}{18} - 5\frac{1}{20}$

15) $16\frac{7}{28} - 5\frac{3}{28}$

16) $5\frac{1}{13} - 2\frac{1}{15}$

17) $51\frac{5}{19} - 4\frac{4}{10}$

18) $28\frac{4}{21} - 4\frac{1}{7}$

19) $16\frac{3}{10} - 2\frac{1}{5}$

20) $14\frac{2}{15} - \frac{1}{16}$

21) $25\frac{1}{8} - 24\frac{1}{9}$

22) $37\frac{2}{12} - 30\frac{2}{13}$

23) $5\frac{6}{7} - 4\frac{1}{21}$

24) $20\frac{3}{13} - 4\frac{2}{26}$

Example:

$$
\begin{array}{rcr}
19\frac{1}{3} & = & 18\frac{6}{3} \\
-\ 2\frac{3}{3} & = & 2\frac{3}{3} \\
\hline
 & & 16\frac{3}{3}
\end{array}
$$

Name ______________________

Date ____________

SUBTRACTION OF FRACTIONS WITH RENAMING

▶ Subtract these fractions. Simplify the answers.

1) $19\frac{1}{5} - 2\frac{3}{5}$

2) $28\frac{2}{7} - 5\frac{5}{7}$

3) $12\frac{2}{17} - 3\frac{5}{17}$

4) $7 - 5\frac{11}{12}$

5) $28\frac{1}{3} - 5\frac{2}{3}$

6) $4\frac{3}{9} - \frac{5}{9}$

7) $6 - \frac{3}{12}$

8) $67 - 5\frac{17}{18}$

9) $13\frac{2}{19} - 2\frac{5}{19}$

10) $6\frac{4}{11} - 1\frac{6}{11}$

11) $8\frac{3}{21} - 5\frac{7}{21}$

12) $8\frac{7}{20} - 5\frac{11}{20}$

13) $14\frac{2}{17} - 5\frac{3}{17}$

14) $29\frac{5}{17} - 28\frac{9}{17}$

15) $32\frac{13}{35} - 5\frac{15}{35}$

16) $5 - 2\frac{17}{20}$

17) $26\frac{5}{28} - 4\frac{6}{28}$

18) $42\frac{16}{50} - 2\frac{25}{50}$

19) $32 - 4\frac{5}{22}$

20) $9\frac{11}{20} - 5\frac{12}{20}$

21) $9\frac{2}{23} - 5\frac{10}{23}$

22) $14\frac{10}{32} - 5\frac{16}{32}$

23) $13\frac{2}{11} - 5\frac{3}{11}$

24) $6\frac{5}{27} - 2\frac{9}{27}$

Example:

$$\begin{array}{rcrcr} 3\frac{2}{5} & = & 3\frac{16}{40} & = & 2\frac{56}{40} \\ -2\frac{7}{8} & = & 2\frac{35}{40} & = & 2\frac{35}{40} \\ \hline & & & & \frac{21}{40} \end{array}$$

Name ____________________

Date ______________

MORE SUBTRACTION OF FRACTIONS WITH RENAMING

▶Subtract these fractions. Simplify the answers.

1) $3\frac{2}{5} - 2\frac{7}{8}$

2) $6\frac{4}{13} - 5\frac{25}{26}$

3) $4 - 2\frac{3}{5}$

4) $9\frac{1}{3} - 4\frac{11}{13}$

5) $14\frac{2}{9} - 5\frac{7}{8}$

6) $16\frac{9}{13} - 4\frac{18}{26}$

7) $3\frac{1}{2} - 2\frac{6}{7}$

8) $12\frac{14}{15} - 10\frac{29}{30}$

9) $11\frac{1}{5} - \frac{1}{4}$

10) $28\frac{1}{15} - 2\frac{3}{4}$

11) $18\frac{2}{15} - 5\frac{3}{4}$

12) $10\frac{2}{5} - 5\frac{10}{11}$

13) $38\frac{2}{11} - 4\frac{5}{22}$

14) $12 - 2\frac{10}{11}$

15) $13\frac{3}{16} - 4\frac{7}{8}$

16) $10\frac{2}{7} - 8\frac{11}{14}$

17) $45\frac{9}{11} - 4\frac{21}{22}$

18) $11\frac{1}{16} - \frac{1}{2}$

19) $23\frac{4}{7} - 22\frac{9}{14}$

20) $4\frac{5}{16} - 2\frac{3}{8}$

21) $7 - 5\frac{2}{15}$

22) $28\frac{2}{17} - 5\frac{9}{10}$

23) $73\frac{3}{4} - 3\frac{16}{18}$

24) $4\frac{5}{13} - \frac{51}{52}$

Example:

$$\begin{array}{r} \frac{7}{18} \\ \frac{3}{-18} \\ \hline \frac{4}{18} = \frac{2}{9} \end{array}$$

Name ______________________

Date ______________

SUBTRACTION OF FRACTIONS

► Subtract these fractions. Simplify the answers to the lowest terms.

1) $\frac{7}{18} - \frac{3}{18}$

2) $2\frac{5}{13} - \frac{4}{13}$

3) $16\frac{2}{7} - 2\frac{3}{7}$

4) $16\frac{3}{10} - 1\frac{5}{10}$

5) $16\frac{12}{13} - 3\frac{1}{26}$

6) $4\frac{15}{16} - 2\frac{3}{8}$

7) $10\frac{13}{24} - 2\frac{1}{6}$

8) $7\frac{5}{11} - 3$

9) $33\frac{1}{5} - 2\frac{3}{4}$

10) $18 - 5\frac{2}{7}$

11) $13 - 2\frac{3}{11}$

12) $16\frac{7}{8} - 3\frac{5}{6}$

13) $59\frac{1}{15} - 2\frac{3}{45}$

14) $23\frac{7}{18} - \frac{3}{72}$

15) $5\frac{1}{10} - 2\frac{7}{15}$

16) $12 - 6\frac{3}{13}$

17) $12\frac{3}{16} - 5\frac{11}{48}$

18) $6\frac{3}{15} - 2$

► Rewrite the fractions in the standard form and subtract. Simplify the answers to the lowest terms.

19) $3\frac{1}{4} - 2\frac{7}{8} =$

20) $13\frac{2}{15} - 3\frac{4}{5} =$

21) $6\frac{5}{12} - 4\frac{1}{5} =$

22) $38 - 2\frac{3}{7} =$

23) $20 - 1\frac{7}{8} =$

24) $30\frac{3}{4} - 16\frac{5}{6} =$

Example:

$2\frac{1}{3} + 3\frac{1}{3} =$

$$\begin{array}{r} 2\frac{1}{3} \\ +3\frac{1}{3} \\ \hline 5\frac{2}{3} \end{array}$$

Name ______________________

Date ______________

BASIC OPERATIONS WITH FRACTIONS AND MIXED NUMBERS

► Add.

1) $2\frac{1}{3} + 3\frac{1}{3} =$ ________

2) $4\frac{1}{8} + 2\frac{2}{8} =$ ________

3) $1\frac{1}{5} + \frac{1}{10} =$ ________

4) $2\frac{1}{6} + 3\frac{2}{3} =$ ________

5) $4\frac{1}{7} + 1\frac{3}{14} =$ ________

6) $1\frac{1}{8} + 2\frac{1}{6} =$ ________

7) $3\frac{2}{3} + 1\frac{1}{9} =$ ________

8) $1\frac{2}{8} + \frac{2}{7} =$ ________

9) $5\frac{3}{9} + 1\frac{1}{2} =$ ________

10) $6\frac{2}{17} + 1\frac{5}{17} =$ ________

► Subtract.

1) $5\frac{2}{8} - 1\frac{1}{4} =$ ________

2) $2\frac{2}{3} - 1\frac{1}{2} =$ ________

3) $4\frac{7}{8} - 1\frac{3}{4} =$ ________

4) $6\frac{2}{8} - 2\frac{3}{4} =$ ________

5) $35 - 6\frac{2}{7} =$ ________

6) $41\frac{3}{5} - 6 =$ ________

7) $21\frac{7}{11} - 4\frac{3}{22} =$ ________

8) $13\frac{1}{5} - 2\frac{3}{6} =$ ________

9) $9\frac{5}{12} - 2\frac{58}{60} =$ ________

10) $4\frac{1}{2} - 2\frac{7}{8} =$ ________

► Multiply.

1) $\frac{5}{6} \times \frac{2}{3} =$ ________

2) $\frac{4}{11} \times \frac{22}{10} =$ ________

3) $\frac{1}{2} \times \frac{2}{3} =$ ________

4) $\frac{3}{5} \times \frac{15}{20} =$ ________

5) $\frac{4}{16} \times 1\frac{1}{2} =$ ________

6) $1\frac{2}{3} \times 2\frac{1}{2} =$ ________

7) $2\frac{3}{5} \times 1\frac{1}{3} =$ ________

8) $6\frac{2}{9} \times \frac{1}{2} =$ ________

9) $1\frac{1}{8} \times 2\frac{3}{5} =$ ________

10) $2\frac{1}{7} \times 1\frac{1}{5} =$ ________

► Divide.

1) $\frac{6}{8} \div \frac{18}{24} =$ ________

2) $\frac{5}{6} \div \frac{25}{30} =$ ________

3) $\frac{4}{11} \div \frac{18}{20} =$ ________

4) $1\frac{1}{2} \div \frac{3}{6} =$ ________

5) $2\frac{1}{4} \div \frac{9}{10} =$ ________

6) $2\frac{3}{5} \div \frac{26}{30} =$ ________

7) $1\frac{2}{9} \div 1\frac{1}{2} =$ ________

8) $3\frac{1}{3} \div \frac{5}{6} =$ ________

9) $1\frac{1}{3} \div 2\frac{3}{4} =$ ________

10) $3\frac{5}{8} \div 3 =$ ________

Name ______________________________

Date ________________

READING AND WRITING NUMERALS

▶ Write the name of the place for each underlined digit.

1) 12.18 tenths
2) 0.920 ____________
3) 1.034 ____________
4) 3.0578 ____________
5) 64.2381 ____________
6) 0.003 ____________
7) 152.9 ____________
8) 24.02231 ____________
9) 8.0263 ____________
10) 4.0005 ____________
11) 4.59021 ____________
12) 295.11 ____________
13) 5.02848 ____________
14) 405.92210 ____________
15) 394.0264 ____________
16) 385.04485 ____________
17) 3.04 ____________
18) 0.003705 ____________
19) 0.00304 ____________
20) 59.0492 ____________
21) 83.39051 ____________

▶ Write the following numerals in words.

22) 9.032 nine and thirty-two thousandths

23) 0.0024 ______________________________

24) 102.10245 ______________________________

25) 0.010139 ______________________________

26) 40.044 ______________________________

27) 410.00003 ______________________________

28) 20.0033 ______________________________

Name ______________________

Date ______________

TRANSLATIONS OF DECIMAL NUMBERS

► Write the following amounts in numerals.

1) Twenty-three and six tenths 23.6
2) Forty-one and three hundredths ______
3) Seventy-two thousandths ______
4) Five and eight tenths ______
5) Six and three thousandths ______
6) One hundred two thousandths ______
7) Four hundred three thousandths ______
8) Two and two hundredths ______
9) Six hundred thirty-four ten-thousandths ______
10) Six thousand, three hundred forty-eight hundred-thousandths ______
11) Twenty thousand, four hundred five hundred-thousandths ______
12) One hundred two and seven hundredths ______
13) Eight hundred two and seven hundred fifty-one thousandths ______
14) One thousand, nine hundred three and seven hundredths ______
15) Two thousand and twenty-six thousandths ______
16) Four thousand three and seven ten-thousandths ______
17) Two hundred six thousandths ______
18) Three hundred-thousandths ______
19) Thirty-four hundred-thousandths ______
20) One and fifty-nine hundredths ______
21) Five ten-thousandths ______
22) Two hundred eleven hundred-thousandths ______

Name ____________________

Date ____________

COMPARING AND ROUNDING DECIMALS

► Arrange each set in order from smallest to largest.

1)	0.6234	62.350	0.7406	0.6234	0.7406	62.350
2)	0.0045	0.0450	0.0040	______	______	______
3)	2.0049	2.0050	2.034	______	______	______
4)	0.1024	0.1031	0.113	______	______	______
5)	23.0045	23.004	2.30045	______	______	______
6)	304.097	300.999	304.102	______	______	______
7)	3.00495	30.0495	0.300495	______	______	______
8)	9.00603	9.00599	9.000999	______	______	______
9)	0.356924	0.350899	0.400001	______	______	______
10)	5.04592	6.001	0.939401	______	______	______

► Round the following numbers to the nearest...

Tenth:	Hundredth:	Thousandth:
11) 245.44 245.4	12) 0.0394 0.04	13) 40.0495 40.050
14) 4.091 ______	15) 199.051 ______	16) 0.08931 ______
17) 2.0399 ______	18) 6.34499 ______	19) 0.00592 ______
20) 0.048 ______	21) 0.995 ______	22) 10.122309 ______
23) 30.9199 ______	24) 666.034 ______	25) 0.39 ______
26) 0.048539 ______	27) 394.091999 ______	28) 390.0485 ______
29) 5.0555 ______	30) 0.0951 ______	31) 3998.0002 ______
32) 49.952 ______	33) 495.0495 ______	34) 8.89099 ______

Name ____________________

Date ____________

Example:
2.3 + 4 + 0.09 + 59 =

```
  2.3
  4.
  0.09
+ 59.
 65.39
```

ADDITION OF DECIMALS

▶ Rewrite the following addends in the vertical form and add.

1) 2.3 + 4 + 0.09 + 59 = ____________

2) 0.056 + 3.02 + 4 + 1.2 = ____________

3) 19 + 9.3 + 0.049 + 3 = ____________

4) 30.9 + 5 + 0.91 + 0.922 = ____________

5) 0.08 + 1 + 1.1 + 6.2 = ____________

6) 2.331 + 0.1123 + 7 + 1.8 = ____________

7) 16 + 4.05 + 5 + 4.77 = ____________

8) 65.94 + 4.7 + 1 + 7.2 = ____________

9) 5.906 + 0.071 + 44.581 = ____________

10) 3.045 + 0.045 + 84.3 = ____________

11) 0.9639 + 0.0082 + 5.03 = ____________

12) 7.304 + 1.5 + 8.33 + 2 = ____________

13) 7004.1 + 35.066 + 0.06 = ____________

14) 93 + 0.739 + 2.38 + 5 + 0.1 = ____________

15) 4.4 + 3.5 + 23.49 + 6 = ____________

16) 6.06 + 33 + 0.045 + 3 = ____________

17) 5 + 6 + 6.9 + 0.082 = ____________

18) 6.112 + 4.7 + 6 + 0.0001 = ____________

19) 9.9 + 5.03 + 5.5 + 0.002 = ____________

20) 47.05 + 6.2 + 0.4 + 1 = ____________

21) 54 + 2.2 + 0.01 + 6.9 = ____________

22) 2.334 + 0.1128 + 8.3 = ____________

23) 4.056 + 3.5 + 7 + 0.92 = ____________

24) 5 + 6.1 + 55.6 + 0.01 = ____________

25) 6.001 + 4.9 + 3 + 0.05 = ____________

26) 1 + 2.1 + 5.66 + 0.031 = ____________

27) 9.9 + 3.4 + 0.56 + 0.012 = ____________

28) 6.0335 + 3.4 + 5 + 0.0567 = ____________

29) 102 + 34.9 + 2 + 30.005 = ____________

30) 5.1023 + 0.7385 + 8 + 3.4 = ____________

31) 26 + 1.5 + 2.153 = ____________

32) 62.43 + 1.6 + 9 = ____________

▶ Solve the following word problems with addition.

33) Compute the total amount deposited if Lance's deposits were $15, $1.90, $121, and $2.65. ________

34) It rained three times during the first week of summer vacation. Compute the total amount if it rained 2.05 inches, 0.29 inches, and 3 inches. ________

Example:

1.03 - 0.94 =

```
 0 9 13
 1.03
-0.94
 0.09
```

Name ____________________

Date ______________

SUBTRACTION OF DECIMALS

▶ Rewrite these problems in the vertical form. Then subtract.

1) 1.03 – 0.94 = ____________

2) From 23.034 subtract 0.0341 ____________

3) 20 – 0.934 = ____________

4) Subtract 12.92 from 27.104 ____________

5) 103.506 – 94 = ____________

6) Subtract 0.607 from 2 ____________

7) 2.3941 – 0.2852 = ____________

8) From 1.0182 subtract 0.81818 ____________

9) 34.8 – 5.0837 = ____________

10) From 2 subtract 1.9283 ____________

11) 0.0238 – 0.003856 = ____________

12) Subtract 9.9 from 10.005 ____________

13) 1.3 – 1.0953 = ____________

14) Subtract 0.0056 from 0.9 ____________

15) 58.3 – 12.923 = ____________

16) From 4.95 subtract 2.5 ____________

17) 71 – 5.341 = ____________

18) Subtract 0.3945 from 6 ____________

19) 0.304 – 0.0433 = ____________

20) From 205.5 subtract 0.56 ____________

21) 4.59 – 2.4 = ____________

22) Subtract 2.3 from 5 ____________

23) 4.5 – 0.0954 = ____________

24) From 6.94 subtract 0.9567 ____________

25) 49 – 5.607 = ____________

26) Subtract 0.0384 from 1.991 ____________

27) 3 – 0.4581 = ____________

28) From 86 subtract 0.86 ____________

29) 7.38 – 0.776 = ____________

30) Subtract 10.9 from 192 ____________

31) 38.76 - 9.8 = ____________

32) From 1 subtract 0.771 ____________

▶ Solve the following word problems with subtraction.

33) Diana saved $25 for school clothes and purchased a blouse costing $9.96. How much money did she have left?

34) Ross drove 298.6 miles on a two-day vacation. If he drove 150 miles on the first day, how many miles did he drive on the second day? ________

Example:

$2{,}300{,}000 = 2.3 \times 10^6$ ← an exponent

a number between one and ten ↖ a power of ten

Name ____________________

Date ______________

SCIENTIFIC NOTATION WITH POSITIVE EXPONENTS

▶ Rewrite the following numbers using scientific notation.

1) 2,300,000 = ________	2) 6,250 = ________
3) 82,100 = ________	4) 50,000 = ________
5) 72,300 = ________	6) 15,080 = ________
7) 1,800 = ________	8) 29,000 = ________
9) 500,000 = ________	10) 600,000 = ________
11) 700,000,000 = ________	12) 7,800,000 = ________
13) 10,000 = ________	14) 35,600 = ________
15) 81.52 = ________	16) 17.63 = ________
17) 236.5 = ________	18) 3,800 = ________
19) 19,000 = ________	20) 16.12 = ________
21) 610,000,000 = ________	22) 400,000,000 = ________
23) 790,000 = ________	24) 25.33 = ________
25) 1,420,000 = ________	26) 1,000,000,000 = ________
27) 34,000,000 = ________	28) 103,000 = ________
29) 23,000 = ________	30) 450,000,000 = ________
31) 11,000 = ________	32) 401,300 = ________
33) 311,400 = ________	34) 102.3 = ________
35) 927,000 = ________	36) 211,400 = ________
37) 100,000 = ________	38) 10,000 = ________
39) 344,000,000,000 = ________	40) 42,000,000,000 = ________
41) 12,000,000 = ________	42) 41,000,000 = ________
43) 764,200,000 = ________	44) 911,400,000 = ________
45) 102,000 = ________	46) 73,000 = ________
47) 21,100 = ________	48) 60,100 = ________

Example:
$.006 = 6 \times 10^{-3}$ ← an exponent
a number between one and ten ↖ a power of ten

Name ____________________

Date ________________

SCIENTIFIC NOTATION WITH NEGATIVE EXPONENTS

▶ Rewrite the following numbers using scientific notation.

1) .006 = ________ 2) .0715 = ________
3) .0062 = ________ 4) .00007 = ________
5) .02 = ________ 6) .0321 = ________
7) .0805 = ________ 8) .0006 = ________
9) .00005 = ________ 10) .03051 = ________
11) .00091 = ________ 12) .0000007 = ________
13) .000003 = ________ 14) .00000021 = ________
15) .0061 = ________ 16) .00054 = ________
17) .000003 = ________ 18) .00101 = ________
19) .000005 = ________ 20) .000052 = ________
21) .000735 = ________ 22) .0001433 = ________
23) .00021 = ________ 24) .00093 = ________
25) .000000004 = ________ 26) .00000000062 = ________
27) .423 = ________ 28) .00316 = ________
29) .005071 = ________ 30) .000078 = ________
31) .002103 = ________ 32) .0000000005 = ________
33) .00000123 = ________ 34) .00203 = ________
35) .000222 = ________ 36) .0121 = ________
37) .10203 = ________ 38) .000204 = ________
39) .0691 = ________ 40) .0000203 = ________
41) .0000000304 = ________ 42) .3044 = ________
43) .00077 = ________ 44) .002058 = ________
45) .0004058 = ________ 46) .03003 = ________
47) .0000000000002 = ________ 48) .00000000000004 = ________

Name ____________________

Date ____________

SCIENTIFIC NOTATION IN STANDARD FORM: Part 1

► Rewrite each scientific notation in standard form.

1)	$5.6 \times 10^2 =$ 560	2)	$1.5 \times 10^2 =$ ______
3)	$2 \times 10^4 =$ ______	4)	$8 \times 10^3 =$ ______
5)	$4.65 \times 10^3 =$ ______	6)	$1.73 \times 10^4 =$ ______
7)	$6.203 \times 10^5 =$ ______	8)	$2.414 \times 10^5 =$ ______
9)	$8.5 \times 10^7 =$ ______	10)	$3 \times 10^7 =$ ______
11)	$2 \times 10^1 =$ ______	12)	$5.6 \times 10^2 =$ ______
13)	$8 \times 10^3 =$ ______	14)	$1.2 \times 10 =$ ______
15)	$7.502 \times 10^2 =$ ______	16)	$3.0052 \times 10^2 =$ ______
17)	$2.61 \times 10^3 =$ ______	18)	$5.85 \times 10^4 =$ ______
19)	$7.05 \times 10^4 =$ ______	20)	$6 \times 10^4 =$ ______
21)	$3.008 \times 10^3 =$ ______	22)	$1.9 \times 10^2 =$ ______
23)	$4.002 \times 10^4 =$ ______	24)	$7 \times 10^{10} =$ ______
25)	$7 \times 10^5 =$ ______	26)	$1.414 \times 10^5 =$ ______
27)	$8 \times 10^1 =$ ______	28)	$3.6 \times 10 =$ ______
29)	$2.5 \times 10^3 =$ ______	30)	$1.4 \times 10^2 =$ ______
31)	$4.5 \times 10^2 =$ ______	32)	$6.6 \times 10^4 =$ ______
33)	$5 \times 10^7 =$ ______	34)	$8.02 \times 10 =$ ______
35)	$7.7 \times 10^6 =$ ______	36)	$2.11 \times 10^6 =$ ______
37)	$5.01 \times 10^7 =$ ______	38)	$1.01 \times 10^8 =$ ______
39)	$1 \times 10^5 =$ ______	40)	$8 \times 10^3 =$ ______
41)	$3.03 \times 10^5 =$ ______	42)	$1.12 \times 10^3 =$ ______
43)	$6 \times 10^9 =$ ______	44)	$9.03 \times 10^2 =$ ______
45)	$8.9901 \times 10^3 =$ ______	46)	$3.2441 \times 10^3 =$ ______
47)	$1.203 \times 10^2 =$ ______	48)	$1.009 \times 10^6 =$ ______

Name ____________________

Date ____________

SCIENTIFIC NOTATION IN STANDARD FORM: Part 2

► Rewrite each scientific notation in standard form.

1) $5.2 \times 10^{-2} =$ .052
2) $3 \times 10^{-3} =$ ________
3) $7 \times 10^{-3} =$ ________
4) $4.6 \times 10^{-2} =$ ________
5) $6.2 \times 10^{-4} =$ ________
6) $2.7 \times 10^{-5} =$ ________
7) $7 \times 10^{-2} =$ ________
8) $8 \times 10^{-4} =$ ________
9) $2.67 \times 10^{-2} =$ ________
10) $7.6 \times 10^{-5} =$ ________
11) $4.83 \times 10^{-4} =$ ________
12) $6.72 \times 10^{-7} =$ ________
13) $8 \times 10^{-6} =$ ________
14) $3.6 \times 10^{-1} =$ ________
15) $7 \times 10^{-7} =$ ________
16) $4.2 \times 10^{-3} =$ ________
17) $4.3 \times 10^{-5} =$ ________
18) $6.21 \times 10^{-4} =$ ________
19) $6 \times 10^{-3} =$ ________
20) $3.01 \times 10^{-2} =$ ________
21) $1.3 \times 10^{-2} =$ ________
22) $4.6 \times 10^{-8} =$ ________
23) $5 \times 10^{-2} =$ ________
24) $1.5 \times 10^{-3} =$ ________
25) $1.9 \times 10^{-8} =$ ________
26) $3 \times 10^{-9} =$ ________
27) $5.03 \times 10^{-1} =$ ________
28) $4.06 \times 10^{-4} =$ ________
29) $6.003 \times 10^{-5} =$ ________
30) $1.01 \times 10^{-5} =$ ________
31) $4.5 \times 10^{-5} =$ ________
32) $5.012 \times 10^{-3} =$ ________
33) $6 \times 10^{-5} =$ ________
34) $7.01 \times 10^{-4} =$ ________
35) $2.34 \times 10^{-3} =$ ________
36) $4.535 \times 10^{-2} =$ ________
37) $1 \times 10^{-8} =$ ________
38) $1.024 \times 10^{-2} =$ ________
39) $4.441 \times 10^{-5} =$ ________
40) $7.002 \times 10^{-4} =$ ________
41) $2.001 \times 10^{-8} =$ ________
42) $3.3 \times 10^{-3} =$ ________
43) $6.77 \times 10^{-4} =$ ________
44) $4.001 \times 10^{-3} =$ ________
45) $5 \times 10^{-10} =$ ________
46) $2 \times 10^{-11} =$ ________
47) $3.42 \times 10^{-6} =$ ________
48) $7 \times 10^{-1} =$ ________

Name ____________________

Date ______________

SCIENTIFIC NOTATION

► Rewrite the following numbers using scientific notation.

1) 2,300,000 = 2.3×10^6

2) 59,000 = ______________

3) 0.0005 = ______________

4) 0.0000039 = ______________

5) 23.41 = ______________

6) 453 = ______________

7) 25,400,000 = ______________

8) 0.000000000843 = ______________

9) 1,900,000,000 = ______________

10) 39,400,000 = ______________

11) 0.00000837 = ______________

12) 567.2 = ______________

13) 0.0001 = ______________

14) 4000 = ______________

15) 0.00495 = ______________

16) 567,000,000,000,000 = ______________

► Write the following numbers in standard form without exponents.

17) 2.3×10^{3} = ______________

18) 4.29×10^{5} = ______________

19) 8×10^{6} = ______________

20) 5.7×10^{5} = ______________

21) 4.94×10^{-8} = ______________

22) 7.03×10^{-7} = ______________

23) 6.1×10^{10} = ______________

24) 5.5×10^{-3} = ______________

25) 6.832×10^{6} = ______________

26) 8.11×10^{-2} = ______________

27) 3×10^{12} = ______________

28) 1.35×10^{-5} = ______________

29) 1.39×10^{7} = ______________

30) 9.04×10^{-4} = ______________

Example:
2.63 x 10 = 2.63
x 10
26.30

Name ____________________

Date ______________

MULTIPLYING DECIMALS BY POWERS OF TEN

▶ Multiply. Write the answer on the line.

1) 2.63 × 10 = ______________

2) 5.638 × 10 = ______________

3) .06 × 100 = ______________

4) .072 × 100 = ______________

5) 1.061 × 10 = ______________

6) 5.63 × 100 = ______________

7) 3.14 × 100 = ______________

8) 1.414 × 1,000 = ______________

9) .00627 × 1,000 = ______________

10) .2802 × 10 = ______________

11) .0605 × 100 = ______________

12) .7701 × 100 = ______________

13) 1.101 × 1,000 = ______________

14) 7.6 × 100 = ______________

15) 5.1 × 1,000 = ______________

16) 8.81 × 10,000 = ______________

17) 3.7 × 10,000 = ______________

18) 2.05 × 10,000 = ______________

19) .0001 × 1,000 = ______________

20) 5.6 × 100 = ______________

21) 69.1 × 1,000 = ______________

22) .777 × 1,000 = ______________

23) .201 × 10,000 = ______________

24) .028 × 10 = ______________

25) .002 × 1,000 = ______________

26) 1.1 × 1,000 = ______________

27) 10 × 1.67 = ______________

28) 1,000 × .003 = ______________

29) 100 × .1505 = ______________

30) 10 × 1.688 = ______________

31) 1,000 × 3.9 = ______________

32) 100 × 3.702 = ______________

33) 10 × .1 = ______________

34) 1,000 × .11 = ______________

35) 3.44 × 100 = ______________

36) 1.112 × 1,000 = ______________

37) .00232 × 10,000 = ______________

38) .012 × 10,000 = ______________

39) 3.033 × 10 = ______________

40) 8.014 × 1,000 = ______________

41) .0556 × 10,000 = ______________

42) 5.5 × 100 = ______________

43) .6709 × 100 = ______________

44) .0021 × 1,000 = ______________

45) 23.1 × 100 = ______________

46) 3.3 × 1,000 = ______________

47) 1.001 × 1,000 = ______________

48) 6.07 × 10,000 = ______________

Example:
$1.2 \times 0.04 =$

$$\begin{array}{r} 1.2 \\ \underline{\times 0.04} \\ 48 \\ 00 \\ \underline{00} \\ 0.048 \end{array}$$

Name ______________________

Date ______________

MULTIPLICATION OF DECIMALS

► Rewrite the following problems in the vertical form and multiply.

1) $1.2 \times 0.04 =$ ______________

2) $48 \times 1.5 =$ ______________

3) $4.05 \times 0.03 =$ ______________

4) $3.6 \times 0.93 =$ ______________

5) $56.7 \times 0.31 =$ ______________

6) $0.059 \times 0.12 =$ ______________

7) $9.01 \times 1.03 =$ ______________

8) $5.8 \times 0.0004 =$ ______________

9) $0.0034 \times 23 =$ ______________

10) $6.12 \times 3.4 =$ ______________

11) $7.81 \times 56 =$ ______________

12) $5.25 \times 0.01 =$ ______________

13) $6.79 \times 8.3 =$ ______________

14) $0.044 \times 0.9 =$ ______________

15) $0.09 \times 0.04 =$ ______________

16) $7.05 \times 0.3 =$ ______________

17) $98 \times 0.11 =$ ______________

18) $0.931 \times 100 =$ ______________

19) $7.02 \times 5.1 =$ ______________

20) $0.034 \times 0.0048 =$ ______________

21) $1000 \times 0.00342 =$ ______________

22) $405 \times 1.52 =$ ______________

23) $8.8 \times 6.7 =$ ______________

24) $13.5 \times 4.7 =$ ______________

25) $69.1 \times 0.001 =$ ______________

26) $10.4 \times 10.5 =$ ______________

27) $0.059 \times 0.0691 =$ ______________

28) $0.101 \times 121.1 =$ ______________

29) $56.9 \times 7.45 =$ ______________

30) $0.005 \times 200 =$ ______________

31) $2.702 \times .06 =$ ______________

32) $0.102 \times .003 =$ ______________

► Solve the following word problems using multiplication.

33) Lionel works part time with a construction company and earns $24.50 per day. How much will Lionel earn working 5 days?

34) Regina earns $7.50 per hour straight time. Compute Regina's time and one-half rate by finding the product of $7.50 and 1.5. ________

Example:

$$\begin{array}{r} 5.3 \\ 10\overline{)53.0} \\ \underline{50} \\ 30 \\ \underline{30} \end{array}$$

Name ______________________

Date ______________

DIVISION OF DECIMALS BY POWERS OF TEN

▶ Divide. Write the answer on the line.

1) 53 ÷ 10 = ____________
2) 7.7 ÷ 100 = ____________
3) .07 ÷ 100 = ____________
4) 39 ÷ 1,000 = ____________
5) 3 ÷ 10 = ____________
6) 4.07 ÷ 100 = ____________
7) 3.02 ÷ 100 = ____________
8) 8.4 ÷ 1,000 = ____________
9) 100 ÷ 1,000 = ____________
10) 5.6 ÷ 10 = ____________
11) 7 ÷ 100 = ____________
12) 6.2 ÷ 1,000 = ____________
13) 1.8 ÷ 1,000 = ____________
14) 94 ÷ 100 = ____________
15) 5 ÷ 1,000 = ____________
16) 13 ÷ 1,000 = ____________
17) 2.6 ÷ 1,000 = ____________
18) 8.6 ÷ 100 = ____________
19) .0023 ÷ 10 = ____________
20) 2 ÷ 1,000 = ____________
21) 3.8 ÷ 100 = ____________
22) 4.02 ÷ 10,000 = ____________
23) 566 ÷ 1,000 = ____________
24) 2,963 ÷ 1,000 = ____________
25) 8,203 ÷ 10,000 = ____________
26) 4,002 ÷ 100 = ____________
27) .706 ÷ 10 = ____________
28) 9 ÷ 1,000 = ____________
29) .04 ÷ 100 = ____________
30) .3006 ÷ 100 = ____________
31) .35 ÷ 1,000 = ____________
32) 4.02 ÷ 100 = ____________
33) 17 ÷ 1,000 = ____________
34) 1 ÷ 1,000 = ____________
35) 4.2 ÷ 10,000 = ____________
36) .02 ÷ 10,000 = ____________
37) 45.7 ÷ 10,000 = ____________
38) 5023.5 ÷ 10,000 = ____________
39) 51.5 ÷ 1,000 = ____________
40) 6.66 ÷ 10,000 = ____________
41) 3.02 ÷ 1,000 = ____________
42) 728 ÷ 10 = ____________
43) 936 ÷ 10,000 = ____________
44) 641.02 ÷ 10,000 = ____________
45) 5.5 ÷ 100 = ____________
46) 1,433 ÷ 100,000 = ____________
47) 2,627 ÷ 10,000 = ____________
48) 300.2 ÷ 100 = ____________

Example:

```
     234
0.1)23.4
    2
    03
     3
     04
      4
```

Name ______________________

Date ______________

DIVISION OF DECIMALS

▶ Rewrite the following division problems in the standard form and divide. Round each quotient to the nearest hundredth.

1) 23.4 ÷ 0.1 = ______________
2) 4.6 ÷ 0.4 = ______________
3) 0.98 ÷ 0.8 = ______________
4) 10 ÷ 5.5 = ______________
5) 1.5 ÷ 0.9 = ______________
6) 2.6 ÷ 1.5 = ______________
7) 0.06 ÷ 0.7 = ______________
8) 40 ÷ 1.2 = ______________
9) 7.7 ÷ 0.03 = ______________
10) 5.6 ÷ 0.12 = ______________
11) 12.3 ÷ 1.1 = ______________
12) 6.99 ÷ 1.2 = ______________
13) 9.12 ÷ 0.9 = ______________
14) 28.04 ÷ 0.7 = ______________
15) 3.5 ÷ 3 = ______________
16) 6.33 ÷ 0.07 = ______________
17) 1 ÷ 0.9 = ______________
18) 2.2 ÷ 13 = ______________
19) 30 ÷ 89 = ______________
20) 5.06 ÷ 1.2 = ______________
21) 5 ÷ 1.5 = ______________
22) 49.9 ÷ 3.4 = ______________
23) 2 ÷ 4.5 = ______________
24) 0.506 ÷ 0.403 = ______________
25) 4.06 ÷ 2.02 = ______________
26) 8 ÷ 0.04 = ______________
27) 0.045 ÷ 0.08 = ______________
28) 39 ÷ 2.6 = ______________
29) 9 ÷ 6.8 = ______________
30) 9.04 ÷ 0.006 = ______________
31) 4.25 ÷ .06 = ______________
32) .913 ÷ 5.5 = ______________

▶ Solve these word problems with division.

33) Kathleen purchased tomatoes for lunch. If the tomatoes were priced at 4 pounds for $4.24, how much did she pay for one pound? ________

34) A carton of six colas sells for $4.62. How much would one cola cost? ________

Name ______________________

Date ______________

BASIC OPERATIONS WITH DECIMALS

▶ Add.

Example: Add

```
  2.3
+ 5.67
 05.90
```

1) 2.3 + 5.67 = ______
2) 45 + 9.97 + .055 = ______
3) 3.04 + .056 + .7 = ______
4) 67.3 + 34.09 + 4.45 = ______
5) .0745 + .45 + .087202 = ______
6) 3.45 + .0923 + 3.07 = ______
7) 2.33 + 0.76 + 74.9 + 4.4 = ______
8) .0348 + .2 + 4 + 4.45 = ______
9) 23.4 + 7.05 + .22 = ______
10) 92 + 8.8 + .403 + .12 = ______

▶ Subtract.

Subtract

```
  5 9 10
  36.00
-   .93
  35.07
```

1) 36 − .93 = ______
2) 4.5 − 2.09 = ______
3) 5.943 − .56 = ______
4) .0982 − .039 = ______
5) 2.9 − .8033 = ______
6) 50.4 − 28.48 = ______
7) 1 − .97 = ______
8) 345 − 23.9 = ______
9) 3.44 − .472 = ______
10) 2.2 − 1.694 = ______

▶ Multiply.

Multiply

```
 1
  2.2
x 0.9
 1.98
```

1) 2.2 × 0.9 = ______
2) 34 × 5.2 = ______
3) 6.7 × .67 = ______
4) 45.3 × .23 = ______
5) 30.5 × 4.5 = ______
6) 3.409 × .42 = ______
7) 90.4 × 2.11 = ______
8) 5.63 × .941 = ______
9) 2.033 × 2.11 = ______
10) 34.6 × 80.4 = ______

▶ Divide. Round off quotients to the thousandths.

Divide

```
   .3822
9)3.4400
  27
   74
   72
    20
    18
     20
     18
      2
```

1) 3.44 ÷ 9 = ______
2) 98.3 ÷ 20 = ______
3) 1.304 ÷ 1.1 = ______
4) 36 ÷ .7 = ______
5) 3 ÷ 1.7 = ______
6) 5.606 ÷ 25 = ______
7) 1.009 ÷ 2.1 = ______
8) 47.03 ÷ 2.7 = ______
9) 1 ÷ 9 = ______
10) 23 ÷ 34 = ______

Example:
Rename $\frac{2}{3}$ as a decimal.

$$\begin{array}{r} .666 \\ 3\overline{)2.000} \\ \underline{1\,8} \\ 20 \\ \underline{18} \\ 20 \\ \underline{18} \\ 2 \end{array} \longrightarrow .67$$

Name ______________________

Date ______________

FRACTIONS TO DECIMALS

▶ Write each fraction as a decimal rounded to the nearest hundredth.

1) $\frac{2}{3}$ = ________ 2) $\frac{3}{16}$ = ________ 3) $\frac{1}{10}$ = ________

4) $\frac{4}{5}$ = ________ 5) $\frac{1}{17}$ = ________ 6) $\frac{13}{14}$ = ________

7) $\frac{6}{9}$ = ________ 8) $\frac{1}{9}$ = ________ 9) $\frac{9}{10}$ = ________

10) $\frac{3}{5}$ = ________ 11) $\frac{5}{14}$ = ________ 12) $\frac{3}{4}$ = ________

13) $\frac{1}{2}$ = ________ 14) $\frac{6}{16}$ = ________ 15) $\frac{11}{12}$ = ________

16) $\frac{3}{11}$ = ________ 17) $\frac{1}{20}$ = ________ 18) $\frac{14}{15}$ = ________

19) $\frac{2}{5}$ = ________ 20) $\frac{5}{21}$ = ________ 21) $\frac{3}{12}$ = ________

22) $\frac{6}{7}$ = ________ 23) $\frac{3}{7}$ = ________ 24) $\frac{1}{11}$ = ________

25) $\frac{1}{8}$ = ________ 26) $\frac{2}{15}$ = ________ 27) $\frac{15}{17}$ = ________

28) $\frac{3}{13}$ = ________ 29) $\frac{3}{16}$ = ________ 30) $\frac{5}{16}$ = ________

31) $\frac{2}{17}$ = ________ 32) $\frac{1}{3}$ = ________ 33) $\frac{2}{19}$ = ________

34) $\frac{1}{6}$ = ________ 35) $\frac{3}{20}$ = ________ 36) $\frac{11}{20}$ = ________

37) $\frac{9}{11}$ = ________ 38) $\frac{6}{19}$ = ________ 39) $\frac{7}{11}$ = ________

40) $\frac{10}{11}$ = ________ 41) $\frac{5}{6}$ = ________ 42) $\frac{7}{16}$ = ________

43) $\frac{5}{13}$ = ________ 44) $\frac{6}{13}$ = ________ 45) $\frac{8}{9}$ = ________

46) $\frac{1}{15}$ = ________ 47) $\frac{1}{4}$ = ________ 48) $\frac{8}{11}$ = ________

Example:
Rename .6 as a fraction.

$.6 = \frac{6}{10} = \frac{3}{5}$

Name ______________________

Date ______________

DECIMALS TO FRACTIONS

► Rewrite each decimal as a fraction. Simplify the answers to the lowest terms.

1) .6 = ______	2) .07 = ______	3) .5 = ______
4) .2 = ______	5) .003 = ______	6) .07 = ______
7) .75 = ______	8) .82 = ______	9) .15 = ______
10) 1.5 = ______	11) .62 = ______	12) .08 = ______
13) .008 = ______	14) .001 = ______	15) 2.6 = ______
16) .022 = ______	17) .04 = ______	18) 20.6 = ______
19) 5.03 = ______	20) 4.1 = ______	21) 200.6 = ______
22) .0012 = ______	23) .041 = ______	24) .101 = ______
25) .9 = ______	26) .102 = ______	27) .052 = ______
28) .0071 = ______	29) .004 = ______	30) .38 = ______
31) 15.1 = ______	32) 3.75 = ______	33) .25 = ______
34) 2.25 = ______	35) 1.82 = ______	36) .21 = ______
37) 1.002 = ______	38) .52 = ______	39) .42 = ______
40) 2.125 = ______	41) .54 = ______	42) .0085 = ______
43) 4.48 = ______	44) 1.53 = ______	45) 10.5 = ______
46) 1.18 = ______	47) 24.5 = ______	48) .0002 = ______
49) .353 = ______	50) 7.2 = ______	51) .0004 = ______
52) .1004 = ______	53) 46.85 = ______	54) .999 = ______
55) .122 = ______	56) .01 = ______	57) .106 = ______
58) .0040 = ______	59) .147 = ______	60) 54.06 = ______
61) 1.85 = ______	62) 9.43 = ______	63) 7.78 = ______
64) 4.14 = ______	65) .335 = ______	66) 1.38 = ______
67) .34 = ______	68) .554 = ______	69) .332 = ______
70) .246 = ______	71) .362 = ______	72) .482 = ______

Example:
Rename .63 as a percent by moving the decimal point two places to the right and adding the percent symbol.

.63 = 63%

Name ______________________

Date ____________

DECIMALS TO PERCENTS

► Rewrite each decimal as a percent.

1) .63 = ______	2) 1.35 = ______	3) .05 = ______
4) .6 = ______	5) .78 = ______	6) .45 = ______
7) .088 = ______	8) .035 = ______	9) .122 = ______
10) .02 = ______	11) .4 = ______	12) .21 = ______
13) 2.09 = ______	14) 2.3 = ______	15) 6.12 = ______
16) 4.5 = ______	17) .065 = ______	18) .0081 = ______
19) .205 = ______	20) .244 = ______	21) .0045 = ______
22) .6031 = ______	23) 5.05 = ______	24) .2207 = ______
25) .41 = ______	26) .2246 = ______	27) .032 = ______
28) .01 = ______	29) .1 = ______	30) .112 = ______
31) 1.72 = ______	32) 1.75 = ______	33) 2 = ______
34) 62 = ______	35) 4.09 = ______	36) 3.1 = ______
37) 9.21 = ______	38) .155 = ______	39) 80 = ______
40) 7.02 = ______	41) 25 = ______	42) 3.44 = ______
43) 22.332 = ______	44) 5.556 = ______	45) 2.33 = ______
46) 75 = ______	47) 15.02 = ______	48) .0062 = ______
49) 30.452 = ______	50) 73 = ______	51) 33.4 = ______
52) 1.433 = ______	53) 43.14 = ______	54) 12.06 = ______
55) 48.045 = ______	56) 2.332 = ______	57) 2.398 = ______
58) 12.1204 = ______	59) 42.46 = ______	60) .0056 = ______
61) 3.449 = ______	62) .000518 = ______	63) .00248 = ______
64) 88 = ______	65) 120 = ______	66) 40.9 = ______
67) 7.881 = ______	68) .01102 = ______	69) .0009 = ______
70) .0104 = ______	71) .33 = ______	72) 14.33 = ______

Example:
8% of 20 is ____.

8% of 20 is N
.08 x 20 = N
1.6 = N

Name ______________________

Date ______________

FIND THE PERCENTAGE

► Solve for the percentage.

1) 8% of 20 is ____________

2) 4% of 53 is ____________

3) 28% of 4 is ____________

4) 6% of 125 is ____________

5) 2% of 86 is ____________

6) 43% of 14 is ____________

7) 75% of 92 is ____________

8) 21% of 34 is ____________

9) 92% of 62 is ____________

10) 53% of 80 is ____________

11) 15% of 28 is ____________

12) 92% of 65 is ____________

13) 3% of 2.1 is ____________

14) 40% of 3.5 is ____________

15) 7% of .7 is ____________

16) 6% of 2.3 is ____________

17) 122% of 42 is ____________

18) 136% of 5 is ____________

19) 200% of 73 is ____________

20) .4% of 96 is ____________

21) 7.2% of 48 is ____________

22) 5.3% of 70 is ____________

23) .8% of 245 is ____________

24) 1.6% of 1.4 is ____________

25) 2.8% of 9.02 is ____________

26) 2.4% of 76 is ____________

27) 5% of .083 is ____________

28) .6% of 435 is ____________

29) .7% of 7.49 is ____________

30) 129% of 4.2 is ____________

31) .03% of 141 is ____________

32) .8% of .2 is ____________

33) .82% of 403 is ____________

34) 245% of 2.6 is ____________

35) 10% of 45 is ____________

36) 3.4% of 500 is ____________

37) 45% of 100 is ____________

38) 23.4% of 300 is ____________

39) 140% of 62 is ____________

40) 7% of 250 is ____________

41) .46% of 746 is ____________

42) 6.5% of 30 is ____________

43) 7.9% of 500 is ____________

44) 200% of .33 is ____________

45) 9% of 9 is ____________

46) 61.3% of 494 is ____________

47) 20% of 10 is ____________

48) 150% of 200 is ____________

Example:
5% of ____ is 1.4

5% of N is 1.4
.05 x N = 1.4
N = 1.4 ÷ .05
N = 28

$$\begin{array}{r} 28 \\ .05\overline{)1.40} \\ \underline{1\,0} \\ 40 \\ \underline{40} \end{array}$$

Name ______________________

Date ______________

FIND THE BASE

▶ Solve for the base.

1) 5% of ________ is 1.4
2) 2% of ________ is 3.6
3) 6% of ________ is 1.8
4) 3% of ________ is .27
5) 7% of ________ is .196
6) 4% of ________ is 1.28
7) 50% of ________ is 2.7
8) 35% of ________ is 30.8
9) 6% of ________ is .078
10) 8% of ________ is .32
11) 2.5% of ________ is .15
12) 7.3% of ________ is 2.044
13) 120% of ________ is 50.4
14) 3.4% of ________ is 2.38
15) 9% of ________ is .315
16) 5% of ________ is .13
17) 5% of ________ is .0235
18) 41% of ________ is .82
19) 3% of ________ is .045
20) 245% of ________ is 191.1
21) 9% of ________ is .0594
22) .3% of ________ is .135
23) .75% of ________ is .9
24) .8% of ________ is .4
25) 4% of ________ is .0176
26) .21% of ________ is .0735
27) 150% of ________ is 135
28) 125% of ________ is 62.5
29) 5.2% of ________ is .0988
30) 3.4% of ________ is .3094
31) .5% of ________ is 1.19
32) .21% of ________ is .0147
33) .32% of ________ is .32
34) .75% of ________ is 7.5
35) 8% of ________ is .416
36) 1.4% of ________ is .196
37) 6% of ________ is .138
38) 5% of ________ is .115
39) 120% of ________ is 102
40) 1% of ________ is .03
41) .9% of ________ is 1.35
42) 6.3% of ________ is .252
43) 8% of ________ is 16
44) 4% of ________ is .024
45) 1.6% of ________ is 3.376
46) .25% of ________ is .0875
47) .4% of ________ is 1.14
48) 1% of ________ is .011

Name ____________________

Date ____________

Example:

____% of 80 is 4.8

```
     6
.80)4.8
    4.8
      0
```

N% of 80 is 4.8

N x .80 = 4.8

N = 4.8 + .80

N = 6%

FIND THE RATE

▶ Solve for the rate.

1) ________% of 80 is 4.8

2) ________% of 35 is 1.05

3) ________% of 70 is 3.5

4) ________% of 30 is 1.8

5) ________% of 80 is 5.6

6) ________% of 20 is 1.8

7) ________% of 80 is .56

8) ________% of 200 is 102

9) ________% of 20 is 7

10) ________% of 300 is 45

11) ________% of 100 is 9

12) ________% of 2,000 is 360

13) ________% of 35 is 2.8

14) ________% of 305 is 48.8

15) ________% of 20 is 30

16) ________% of 64 is 112

17) ________% of 10.5 is .84

18) ________% of 250 is 70

19) ________% of 38 is 1.14

20) ________% of 72 is 1.08

21) ________% of 206 is 16.48

22) ________% of 500 is 4.5

23) ________% of 2.06 is .1442

24) ________% of 600 is 420

25) ________% of 700 is 9.8

26) ________% of 40 is .32

27) ________% of 200 is .14

28) ________% of 3,000 is 1.8

29) ________% of 2.5 is .9

30) ________% of 4.2 is .546

31) ________% of 4 is .12

32) ________% of 400 is .08

33) ________% of 22 is 56.1

34) ________% of 30 is 106.5

35) ________% of 80 is 14

36) ________% of 80 is 5.6

37) ________% of 90 is 5.4

38) ________% of 50 is 100

39) ________% of 23 is 2.3

40) ________% of 120 is 300

41) ________% of 22 is .33

42) ________% of 30 is 2.85

43) ________% of 70 is 4.9

44) ________% of .16 is .000136

45) ________% of 50 is 9

46) ________% of 70 is .21

47) ________% of 1.5 is .018

48) ________% of 200 is .14

Name ______________________

Date ______________

PERCENT SENTENCES

▶ Solve for the percentage.

1) 25% of 60 is __15__ .
2) 82% of 50 is ______ .
3) 90% of 60 is ______ .
4) 35% of 36 is ______ .
5) 92% of 100 is ______ .
6) 20% of 30 is ______
7) 80% of 90 is ______ .
8) 23% of 65 is ______ .
9) 10% of 150 is ______ .
10) 9% of 250 is ______ .

▶ Solve for the base.

1) 15% of __60__ is 9.
2) 60% of ______ is 15.
3) 20% of ______ is 8.
4) 62% of ______ is 10.54.
5) 53% of ______ is 106.
6) 92% of ______ is 23.
7) 17% of ______ is 51.
8) 30% of ______ is 21.
9) 62% of ______ is 19.84.
10) 35% of ______ is 21.

▶ Solve for the rate.

1) __80__% of 9.5 is 7.6.
2) ______% of 70 is 21.
3) ______% of 45 is 4.05.
4) ______% of 30 is 21.
5) ______% of 80 is 72.
6) ______% of 26 is 0.65.
7) ______% of 36 is 1.98.
8) ______% of 30 is 9.
9) ______% of 70 is 10.5.
10) ______% of 50 is 65.

▶ Complete each percent sentence.

1) __15__% of 2.5 is 0.375.
2) 5.5% of _____ is 0.2475.
3) 0.5% of 75 is ______ .
4) 2.8% of _____ is 18.2.
5) _____% of 240 is 192.
6) 75% of ______ is 225.
7) 10% of ______ is 0.26.
8) _____% of 6.2 is 1.24.
9) 0.8% of 80 is ______ .
10) 0.25% of 16 is ______ .

Example:

```
  5 weeks  3 days
+          4 days
  5 weeks  7 days  =  6 weeks
```

Name ____________________

Date ______________

UNITS OF TIME

▶ Perform the indicated operations. Simplify the answers.

```
1)    5 weeks 3 days            2)    3 years 2 weeks 6 days
   +          4 days               +  2 years 6 weeks 3 days
```

```
3)    7 weeks 4 days            4)    9 weeks 5 days
   +  2 weeks 5 days               −  7 weeks 6 days
```

```
5)    8 years 6 weeks 3 days    6)    5 weeks 3 days
   −  3 years 5 weeks 6 days       −          6 days
```

```
7)    15 years 9 weeks          8)    10 years 2 weeks 1 day
   −   7 years 7 weeks             −   8 years 5 weeks 4 days
```

```
9)    3 days 12 hours 10 min.   10)   5 hours 10 min. 5  sec.
   +  3 days  7 hours 56 min.      +  8 hours 50 min. 58 sec.
```

```
11)   2 hours 20 min. 4 sec.    12)   4 hours 10 min.
   −  1 hour  10 min. 7 sec.       −  3 hours  9 min. 5 sec.
```

```
13)   3 hours 2 min.            14)   7 hours  8 sec.
   −  2 hours 8 min.               −  6 hours 10 min.
```

15) 3 days = ________ hours

16) 18 minutes = ________ seconds

17) 3 years = ______ weeks

18) 180 seconds = _______ minutes

19) 5 weeks 3 days − 3 weeks 5 days 3 hours = ____________________

20) 8 years 8 weeks 2 days + 2 years 9 weeks 5 days = ______________

21) 4 hours 3 min. 34 sec. − 2 hours 7 min. 50 sec. = _______________

22) 5 min. 45 sec. + 5 hours 4 min. 49 sec. = ____________________

23) 35 hours 24 sec. − 9 hours 28 min. = ________________________

24) 55 min. 19 sec. + 1 hour 20 min. 4 sec. = ____________________

25) 36 hours − 3 hours 5 min. 6 sec. = __________________________

Example:
Find the area of a triangle whose base is 23 feet and height is 8 feet.

Area = $\frac{1}{2}$ base x height
A = $\frac{1}{2}$ (23) x 8
A = 11.5 x 8
A = 92 square feet

Name ______________________

Date ______________

AREA OF A TRIANGLE

► Find the area of each triangle described below.

1) Base = 23 feet
Height = 8 feet ____________

2) Base = 41 feet
Height = 32 feet ____________

3) Base = 40 feet
Height = 10 feet ____________

4) Base = 50 feet
Height = 12 feet ____________

5) Base = 12 feet
Height = 10 feet ____________

6) Base = 5 feet
Height = 4 feet ____________

7) Base = 20 inches
Height = 10 inches ____________

8) Base = 33 inches
Height = 50 inches ____________

9) Base = 9 meters
Height = 18 meters ____________

10) Base = 10 meters
Height = 15 meters ____________

11) Base = 6 cm
Height = 17 cm ____________

12) Base = 3 meters
Height = 20 meters ____________

13) Base = $\frac{2}{3}$ inch
Height = 12 inches ____________

14) Base = $\frac{3}{4}$ foot
Height = 16 feet ____________

15) Base = 1.2 cm
Height = 6 cm ____________

16) Base = 10 meters
Height = 4.5 meters ____________

17) Base = 20 yards
Height = 5 yards ____________

18) Base = 10 feet
Height = 2 feet ____________

19) Base = .5 cm
Height = .2 cm ____________

20) Base = 1.1 cm
Height = .5 cm ____________

21) Base = $\frac{7}{8}$ inch
Height = $\frac{1}{2}$ inch ____________

22) Base = $\frac{4}{5}$ inch
Height = $\frac{1}{2}$ inch ____________

Example:
Find the area and circumference of a circle with a radius of 5 inches.

Area $= \pi r^2$
$= 3.14 \times 5^2$
$= 3.14 \times 25$
$= 78.50$ square inches

Circumference $= 2 \pi r$
$= 2 \times 3.14 \times 5$
$= 31.4$ inches

Name ____________________

Date ________________

AREA AND CIRCUMFERENCE OF CIRCLES

► Solve for the area and circumference for each of the circles described below. Use 3.14 for pi. The abbreviation for *circumference* is C.

1)
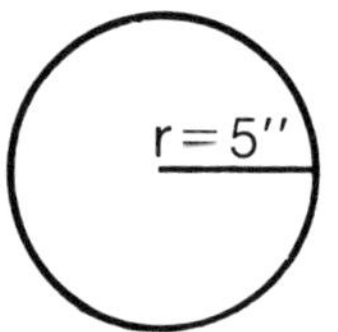

Area = __________
C = __________

2)
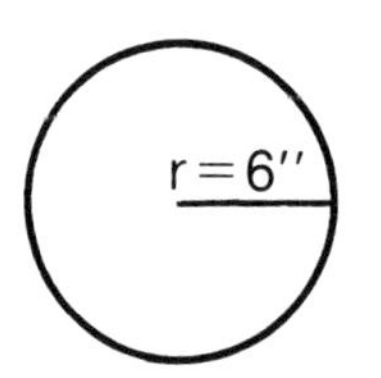

Area = __________
C = __________

3)
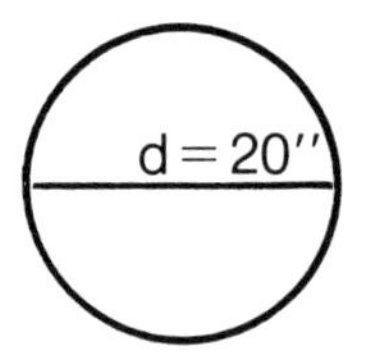

Area = __________
C = __________

4)
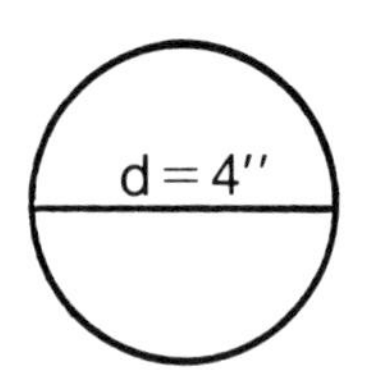

Area = __________
C = __________

5)
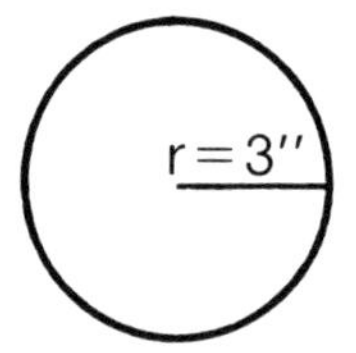

Area = __________
C = __________

6)
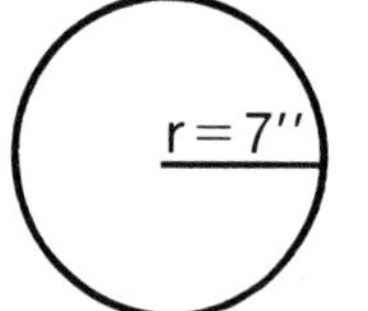

Area = __________
C = __________

7)
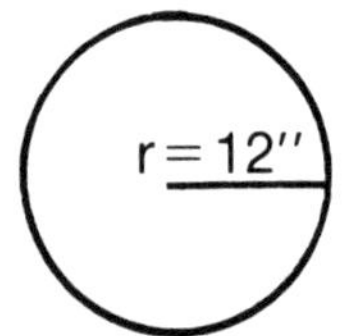

Area = __________
C = __________

8)
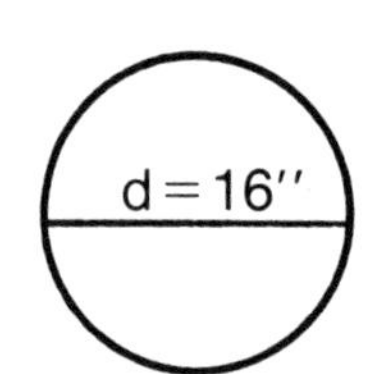

Area = __________
C = __________

9)
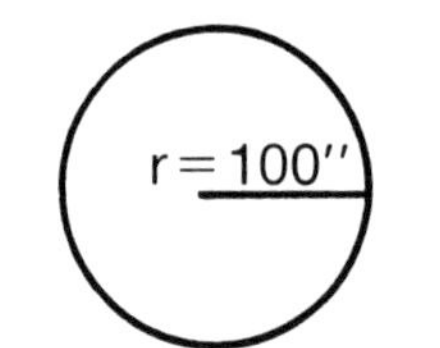

Area = __________
C = __________

10)
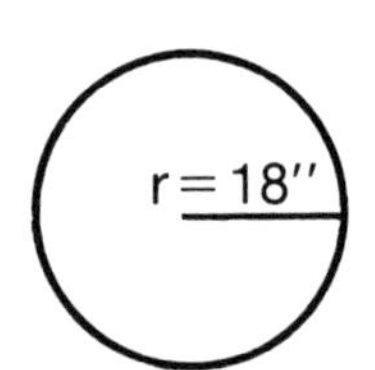

Area = __________
C = __________

11)
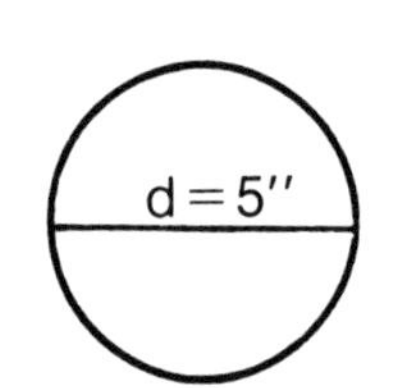

Area = __________
C = __________

12)
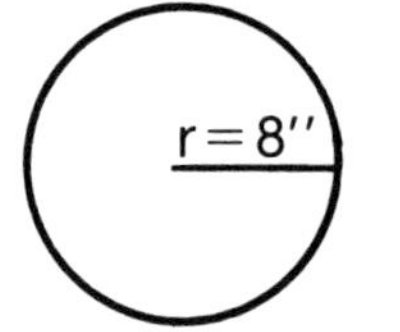

Area = __________
C = __________

13) Radius = 10 in.
Area = ________
C = ________

14) Radius = 20 cm
Area = ________
C = ________

15) Radius = 5 cm
Area = ________
C = ________

16) D = 20 meters
Area = ________
C = ________

17) D = 1.4 cm
Area = ________
C = ________

18) D = 100 mm
Area = ________
C = ________

Name ____________________

Date ____________

MEASURING ANGLES

► Use a protractor to measure these angles to the nearest degree.

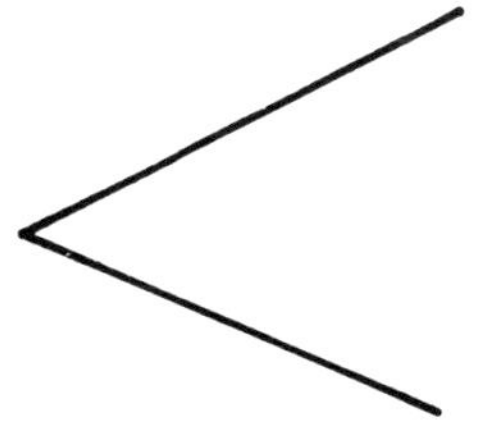

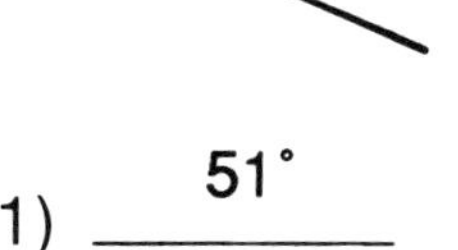

1) 51°

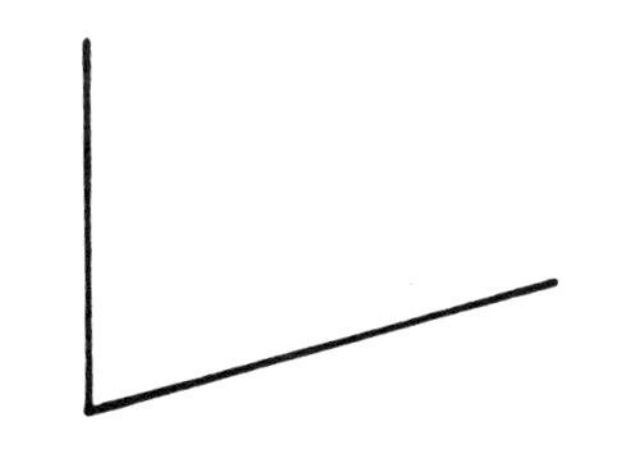

2) ______°

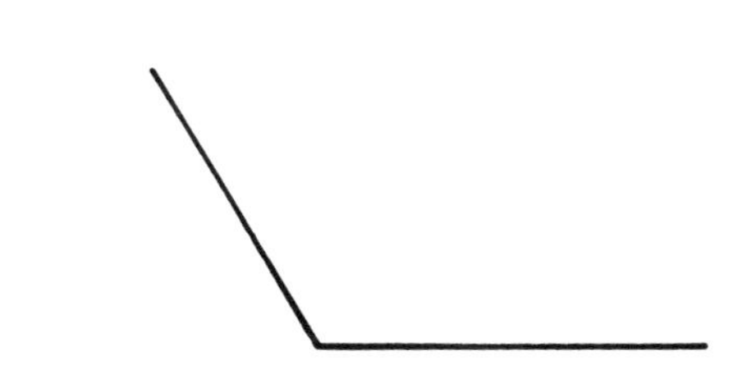

3) ______°

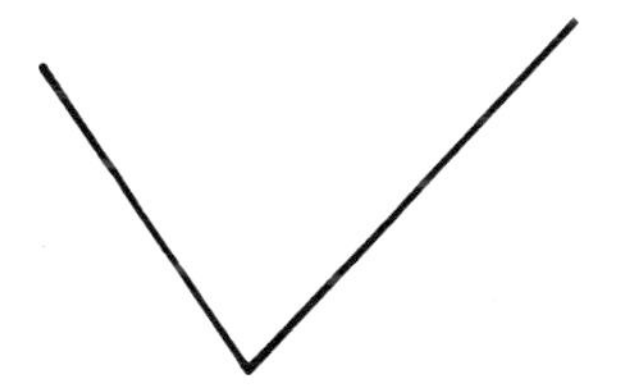

4) ______°

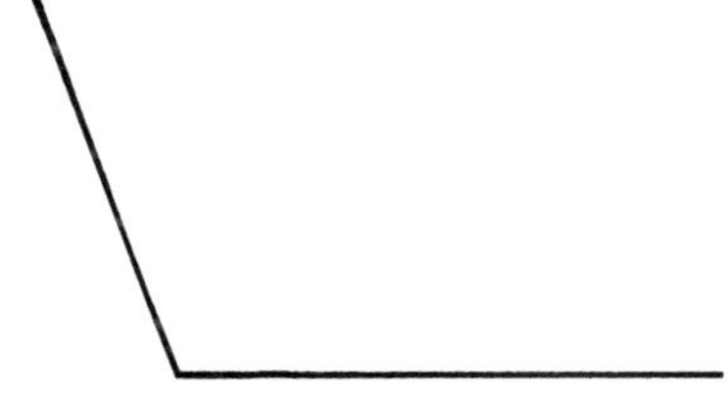

5) ______°

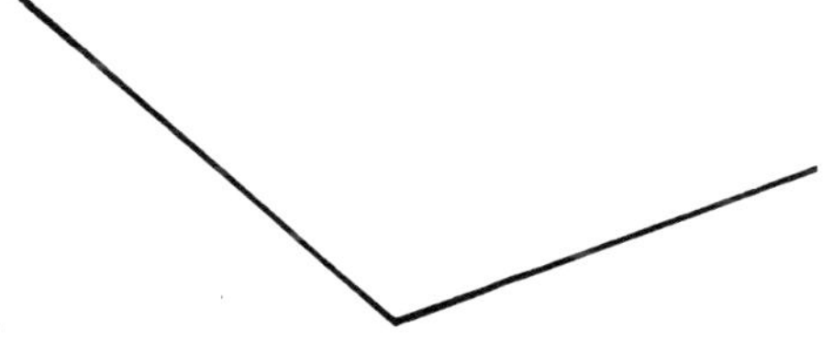

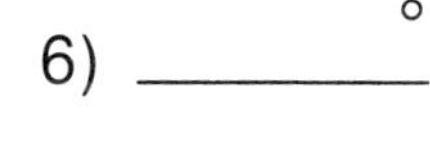

6) ______°

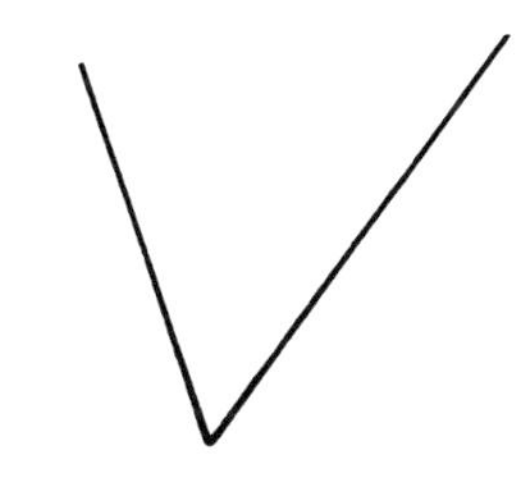

7) ______°

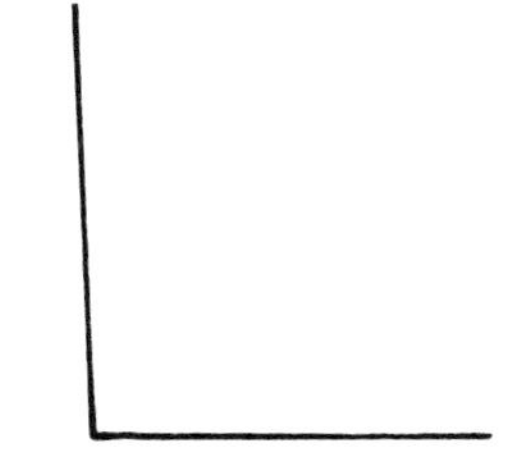

8) ______°

9) ______°

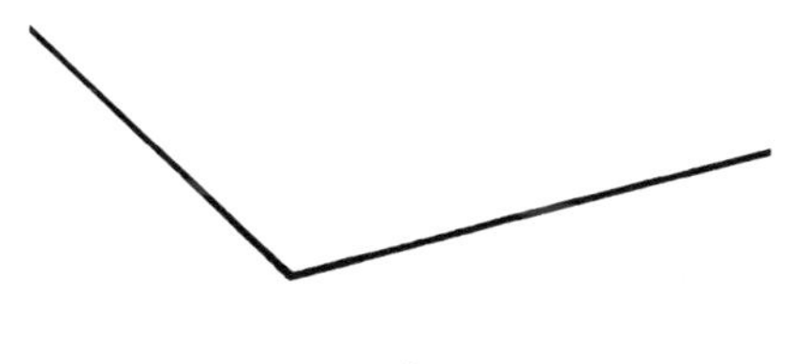

10) ______°

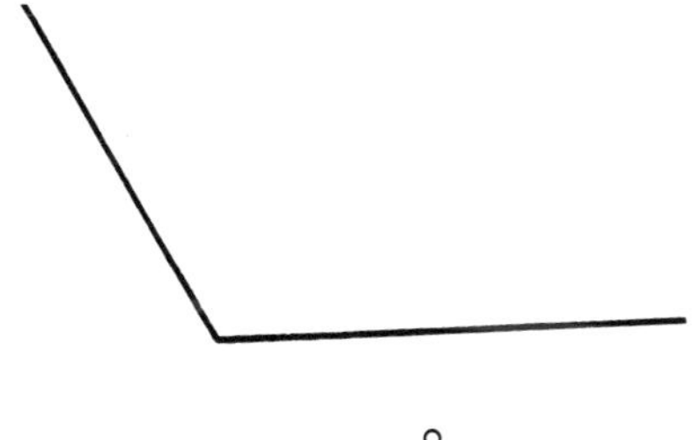

11) ______°

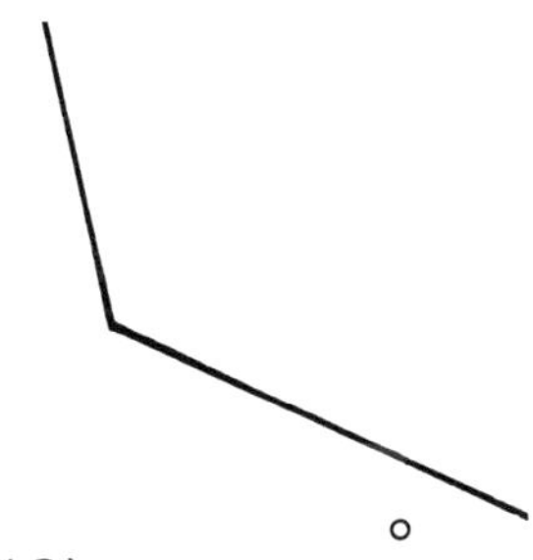

12) ______°

Name ____________________

Date ____________

USING THE METRIC SYSTEM

► Measure these line segments using the metric system. Measure to the nearest millimeter. Write your answer on the line.

1) 5 cm 5 mm ____________ 2) ____________

3) ____________ 4) ____________ 5) ____________

6) ______ 7) ____________ 8) ____________

9) ______ 10) ____ 11) ________ 12) ____________

13) ______________________________ 14) ______

15) ________ 16) ______________________________

17) ____

18) ____________________ 19) ____________________

20) ______________ 21) ________________________

22) _____ 23) ___ 24) ____________________________

25) __________________________ 26) ____________

27) ________ 28) ______________________________

29) __________________________________ 30) __________

31) _____ 32) __ 33) __________ 34) ______________

35) __________________________ 36) ____________________

37) ________ 38) ____________ 39) ______________

40) ____